Why ORP Systems Fail - and How to Stop it

A Practical Guide for Trained Professionals

David Watson

The swimming pool bloke who reads a lot

Publisher: David Watson Pool Consulting

Published by: David Watson Pool Consulting
www.davidwatsonpoolconsulting.com

Design and illustrations: Sue Knox
Cover Design: David Watson

Ingram Spark / Amazon KDP Print-on-Demand Edition
ISBN EBook: 978-1-7645296-2-4
ISBN Paperback: 978-1-7645296-3-1

Printed in 2026

In preparing this handbook, AI tools have assisted with editing and verifying references across multiple international databases.

All content has been carefully reviewed and edited for accuracy and reliability, informed by my four decades of field experience and current regulatory standards.

Acknowledgement

I extend my sincere thanks to the early readers who reviewed the initial drafts of this book.

Your thoughtful feedback, careful reading, and constructive suggestions made this work stronger and clearer.

I am grateful for your support in developing this manuscript.

Contents

A pool is not just water.
It is a series of interrelated systems.

It is our responsibility.
We own every drop until it's safe.
When a bather walks out the gate,
they carry our standard with them.
They take it home, to work,
to their communities.

Their safety —
and the safety of everyone they touch —
rests with us.

CHAPTER 1

Special Mention

Acknowledgement of contributions that influenced the technical accuracy of this book.

Many people help with writing a book like this. They are usually working quietly in the background — perhaps a beta reader who makes a few comments, or someone who has a conversation over coffee about something tangentially related.

Over many years, many people have mentioned things that helped shape the ideas expressed in this book.

However, one contribution deserves special recognition:

I would like to thank Richard A. Falk for his generous and detailed technical feedback on chlorine chemistry, redox behaviour, and ORP interpretation. His work clarifying hypochlorous acid / hypochlorite ion ($HOCl/OCl^-$) kinetics and thermodynamics significantly strengthened the accuracy of this book.

Richard has helped the pool industry — and me — more than can be expressed in a simple acknowledgement. Not only with this book, but in so many ways.

Thank you, Richard. Fantastic work.

CHAPTER 2

What This is About

This book explains why ORP control behaves unpredictably in real pools and how to understand what is actually happening.

Have you ever encountered a situation like this?

There is a spa or pool installed on your site.

It was testing fine. The controller was reading correctly on the last visit.

Two days later, you receive a call reporting excessive chlorine in the water, or no chlorine at all, because the dosing system has emptied the drum completely.

You attend the site and find the full drum of chlorine, acid, or both has been fed in.

Yet the readings were perfect a few days before.

This becomes even more confusing when the spa has been stable for several weeks, months, or sometimes years.

Or consider this scenario:

The water tested fine.

The controller read correctly.

Calibration[1] was completed and set points[2] all ticked off this morning.

---

1 Adjusting a measuring device so it shows the correct value.

2 The target value you want a system to reach or maintain.

Then this afternoon, chlorine was over 30 mg/L, and swimmers' swimwear was being bleached.

Events like these can lead to:

- Environmental Health Officers making statements like, "No spray pads will be permitted in this city."
- Facilities being shut down and improvement notices issued.
- Damage to equipment, bathers, and potentially you.

The reason these events happen is not equipment failure. The

controller was not broken. It was doing exactly what it does. The problem is that most of what the industry teaches about ORP (the millivolt reading from the probe) does not explain what that reading actually means.

After forty years of working with these controllers since about 1988, I have enough experience to be genuinely sceptical of them, and enough respect for the technology to have written this book anyway.

> **Note:** *This book assumes you already understand:*
>
> - *free chlorine testing*
> - *pH testing*
> - *probe calibration*
> - *total alkalinity basics*
> - *circulation systems*
>
> *This book also presupposes that the reader already understands dosing systems and controller calibration procedures and focuses instead on how ORP behaves inside real operating systems.*

This book explains how controllers interpret water conditions. The

optional companion CO_2 book explains the carbonate system that drives those conditions.

CHAPTER 3

How ORP Really Works (Read This First)

Before going further, understand this:

A pool is a system, not a tank of water.

ORP does not control the pool.

It only reacts to what the system is doing.

The chain works like this:

Hydraulics → Mixing → Chemistry → pH → ORP → Controller

If anything, earlier in that chain is unstable.

Everything after it will also be unstable.

Core rules:

- ORP measures electron activity related to chlorine but is not a direct chlorine measurement.
- ORP follows pH
- pH follows CO_2
- CO_2 follows how the water is moved and disturbed

This means:

If CO_2 is unstable, → pH will move

If pH moves → ORP will move

If ORP moves → the controller will respond

If these are not stable, the controller cannot be stable.

In addition, if the system is not stable, the ORP controller cannot make it stable.

The controller is reacting to the problem.

It is not creating stability.

In addition, if the system itself is not stable, the ORP controller cannot make it stable.

This book explains each part of the chain.

CHAPTER 4

Origins

This chapter explains my practical background with ORP controllers and why I approach them with caution.

I have been involved in recreational water for a while — over forty years now, keeping bathers safe and preventing recreational water illnesses wherever I can.

Back in the late 1980s, around 1988, I was selling and installing Hannah Instruments Blackstone ORP/pH controllers, as early pool-specific controllers had not appeared in my local market.

The Blackstone controllers were semi-industrial units, and I was installing them on both domestic and commercial pools.

So, when I talk about ORP/pH, it is not from the outside looking in. I have been working with this technology long before it became widely adopted.

Even after all that experience, I have always been careful about how ORP/pH controllers are used in recreational water.

They are just one component in an entire network of systems that keep water safe. The sanitiser, whether chlorine, bromine, or another agent, is the last line of defence.

ORP/pH controllers are only part of the system, not the hero of the story, regardless of what the marketing often suggests.

I have always been cautious about relying on them in pools.

In the industrial and drinking-water world, where I did a fair amount of ORP/pH work, they behave more predictably.

In recreational water, they have always struck me as too unreliable, too sensitive, too easy to mis-set, and too often sold as a magic fix when they are not.

I have also included a section on bulk chemical storage in this book. That comes from another chapter of my life: I was a dangerous goods consultant in Western Australia.

When you turn up on site, and there is a large tank of chlorine or acid in front of you, it helps to have even a basic understanding of what you are dealing with.

So, I have included that background for readers who may walk into these situations.

This book was not my idea. Lee Salisbury from the Institute of Research and Learning asked me to write it.

Lee reviewed earlier drafts of my other books, so when she asked, I agreed — part curiosity, part "oh great, here we go."

Perhaps I am wrong about ORP/pH.

Perhaps there is a way to make it more reliable in recreational water.

Perhaps, with a better understanding than what is found in most manuals, the problems can be overcome.

One thing I want to make clear:

I do not see ORP/pH control as a "chemical system."

I see it as one interlinked subsystem inside a recreational water body, connected to hydraulics, circulation, design, bather load, filtration, chemistry, and everything else that affects water quality.

So, this book is part research, part observation, and part me revisiting decades of experience to see whether new insights emerge when everything is laid out together.

ORP/pH is likely one of the more misunderstood technologies in the industry.

This is going to be an interesting journey — for me as much as for the reader.

Because of my work and confidentiality agreements in place, detailed case studies and specific client examples are not included in this book. Instead, the guidance and insights provided are based on extensive experience and best practices distilled from a wide range of projects.

CHAPTER 5

What This Book Is / Isn't

Defines what this book will help you do and, just as importantly, what it will not solve.

This book provides a clear, practical, and field-focused understanding of ORP/pH control.

ORP is not unreliable. It is conditional.

No chemical controller stabilises a pool. Hydraulics, turnover, mixing, buffering, and bather load management create stability.

Controllers only respond to the stability already present in the system.

Rule: Controllers do not create stability. They only respond to it.

If the system is unstable, the controller will just follow it into the ditch.

This book assumes the reader is familiar with swimming pool or spa pool equipment.

A note for early-career technicians

At this stage in your career, your role is not to solve ORP. It is to recognise when ORP cannot be trusted.

Purpose of the Book:

- Explain ORP and pH control in clear, service person friendly language.
- Replace industry folklore with real, evidence-based understanding.
- Help operators diagnose and prevent common ORP/pH failures.

Problems This Book Solves:

- Misunderstanding of ORP readings and their relationship to pH and CO_2.
- Incorrect installation or placement of the probes.
- Instability caused by hydraulics, aeration, temperature, or contaminants.
- Overshooting, undershooting, and controller misbehaviour.

What This Book Will Not Solve:

- It does not turn ORP into a perfect or universal control method.
- It does not fix inappropriate pool designs, poor circulation, or bad hydraulics.
- It does not replace proper operator training or regulatory requirements.
- It cannot overcome the inherent limits of ORP in spas, spray pads, or highly aerated pools.

Who This Book Is For:

- Technicians, operators, and service personnel who manage real-world pools.
- Facility managers seeking to understand ORP behaviour.
- People transitioning from manual dosing to automated systems.
- Trainers and supervisors wanting a reliable technical reference.

Who This Book Is Not For:

- Readers seeking academic electrochemistry or advanced modelling.
- Pools relying solely on amperometric or colorimetric systems.
- Water bodies unsuitable for ORP control (spas, spray pads, high-aeration designs).
- Anyone expecting ORP to replace consistent manual testing.

How This Book Is Built

This book follows one chain, in order:

Hydraulics → Mixing → Chemistry → pH → ORP → Controller

Each stage depends on the one before it. If you get the order wrong in your thinking, the rest of the book will not make sense.

So the book moves in four stages:

1. **What's actually in the water?**
   Before ORP means anything, you need to know what is moving through the system and what is reacting in it: hydraulics, turnover, the chemicals themselves.

2. **How ORP behaves, and why it is not what you think.**
   This is the physics and chemistry of the reading itself: what raises it, what suppresses it, why it follows pH rather than chlorine.

3. **Building a system that does not fight itself.**
   Site setup, storage, tubing, pumps, injectors: the physical installation that determines whether the controller has a chance of working at all.

4. **Stage four — Keeping it honest over time.**
   Probes drift. Chemicals degrade. Storage temperature matters. This is the maintenance and troubleshooting half.

If you are already experienced in the field, you can skip to Stage 3 or 4 and use the earlier chapters as a reference. If you are newer to this work, read in order, each stage builds on the last.

CHAPTER 6

Safety Information

Sets the boundaries for safe handling of chemicals and equipment when applying anything in this book.

The information in this book is provided for educational and informational purposes only.

While every effort has been made to ensure accuracy, the author and publisher assume no responsibility or liability for any errors or omissions, or for any damages or injuries resulting from the application of the information.

Handling pool chemicals and equipment involves inherent risks.

Always follow manufacturer instructions, use personal protective equipment (PPE), and comply with all applicable local, regional, and national laws and regulations.

This book is not a substitute for professional training, certification, or legal advice.

Readers are encouraged to consult qualified professionals for help with pool installation, maintenance, chemical dosing, or any safety concerns.

By using this book, you agree to do so at your own risk and acknowledge that the author and publisher are not liable for any consequences arising from its use.

CHAPTER 7

Glossary

Aeration

The process of introducing air into the water, which affects pH and ORP readings because of oxygen content and CO_2 exchange.

Alkalinity

The measure of the water's ability to neutralize acids, primarily because of bicarbonate ions, affects pH stability.

Amperometric Sensor

A sensor type that measures the concentration of free chlorine directly by detecting the current generated by the chemical reaction.

Bromine

A chemical sanitiser used in pools and spas, less common than chlorine, with different ORP characteristics.

Cavitation

Formation and collapse of vapour bubbles in a liquid near pumps or valves, causing turbulence and damaging equipment.

Cyanuric Acid (CYA)

A stabiliser that protects chlorine from UV degradation but reduces the fraction of active hypochlorous acid.

Dosing Pump

A device that injects chemicals like chlorine or acid into the pool water in controlled amounts.

Electrode (ORP/pH Probe)

A sensor that measures oxidation-reduction potential or pH by detecting electrical signals resulting from chemical reactions.

Free Chlorine

The amount of chlorine available in the water to sanitise and oxidise contaminants, including hypochlorous acid and hypochlorite ion.

Hydraulics

The study and management of water flow and circulation within the pool system, important for consistent chemical distribution.

Hypochlorous Acid (HOCl)

The active, fast-acting form of chlorine responsible for sanitizing pool water.

Nernst Equation

A mathematical relationship describing how ORP changes with ion concentration and temperature.

Oxidation-Reduction Potential (ORP)

A measure of the water's ability to oxidise contaminants, expressed in millivolts (mV), reflecting the electron activity in the water.

Peristaltic Pump

A dosing pump type that uses rotating rollers to compress tubing, providing precise chemical dosing.

pH

A scale that measures the acidity or alkalinity of water, critical in pool chemistry and ORP interpretation.

Polymers

Large molecules used in some water clarifiers and algaecides that can foul ORP and pH probes.

Redox Reaction

Chemical reactions involving the transfer of electrons between substances, fundamental to ORP measurement.

Salt Chlorinator

A device that generates chlorine from dissolved salt (sodium chloride) in pool water via electrolysis.

Setpoint

The target value you want a system to reach or maintain. This refers in pools to either the pH target or the ORP target.

Total Dissolved Solids (TDS)

The combined content of all inorganic and organic substances dissolved in water.

Turnover

The time it takes for the entire volume of pool water to pass through the filtration system once.

CHAPTER 8

What are we Controlling

Explains that a pool is a system of interacting factors, and ORP is only one part of it.

A recreational water body is a semi-open (a system affected by outside conditions) environmental control system.

It is influenced by:

- hydraulics
- atmospheric exchange
- human loading
- biochemistry
- surface chemistry
- temperature
- materials
- and instrumentation limits

These factors interact to determine the condition of the water.

ORP controllers operate within this system.

They are influenced by it.

They do not control it.

Rule: ORP does not control the pool. The system controls ORP.

This means:

- Changes in hydraulics, loading, or environment will change chemistry.
- Changes in chemistry will change the pH.
- Changes in pH will change ORP.

The controller responds to that change.

It does not prevent it.

These interacting parts will be examined further in this book.

To understand ORP/pH, each part of the system must be considered as part of the whole.

ORP as a System Indicator (Not *a Chlorine Meter)*

Early chapters explaining "What are we controlling"

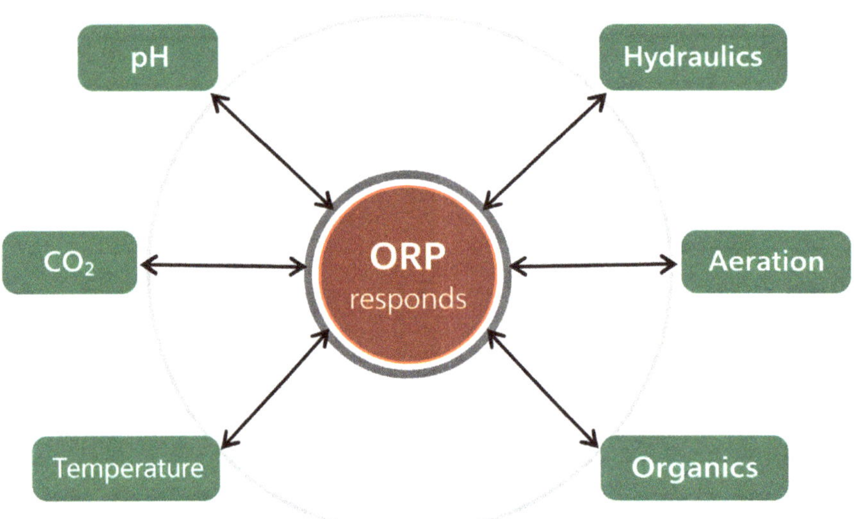

ORP responds, it does not command.

CHAPTER 9

Hydraulics and Turnover

Shows how water movement, mixing, and flow conditions directly affect ORP behaviour.

Why this matters:

Poor mixing creates unstable chemistry.

Unstable chemistry creates unstable ORP.

> **Rule:** Poor mixing = unstable chemistry = unstable ORP.

About

Before commencing on the chief topic of ORP/pH, there is a need to quickly cover a few areas in hydraulics and turnover.

This is not a complete treatment of hydraulics.

It is a focused look at what directly affects ORP/pH behaviour.

It is assumed that the reader already has a basic working knowledge of these topics.

Turnover vs Mixing

Turnover expresses the time it takes for the pool volume to pass once through the entire filtration process.

Typically, turnover is expressed in hours.

A 4–6 hour turnover is a common domestic turnover rate. 2.5 hours is a common commercial turnover rate in WA.

However:

One turnover does not mean all the water has been treated.

It does not mean all the water has passed the probe.

What is actually happening:

A water body works on the principle of continuous dilution (Water is mixed like stirring milk into tea repeatedly, not replaced all at once.)

This means:

Some water is treated multiple times.

Some water is treated for the first time.

Some water has not yet reached the treatment system.

Result:

Treatment is uneven.

It depends entirely on how well the water is mixed.

Under the theory of continuous dilution:

Roughly six turnovers are required for >95% probabilistic treatment exposure.

Practical implication:

Most times, one turnover is sufficient for basic chemical testing.

However, this is not always the case.

Why this varies:

Different water bodies mix at different rates.

A standard rectangular pool will generally mix well.

A free-form pool with curves, nooks, and corners will often have reduced circulation and poorer mixing.

What this means for ORP:

If the water is not evenly mixed:

→ Chemistry varies across the pool

→ The probe sees only part of the system

→ ORP readings become inconsistent.

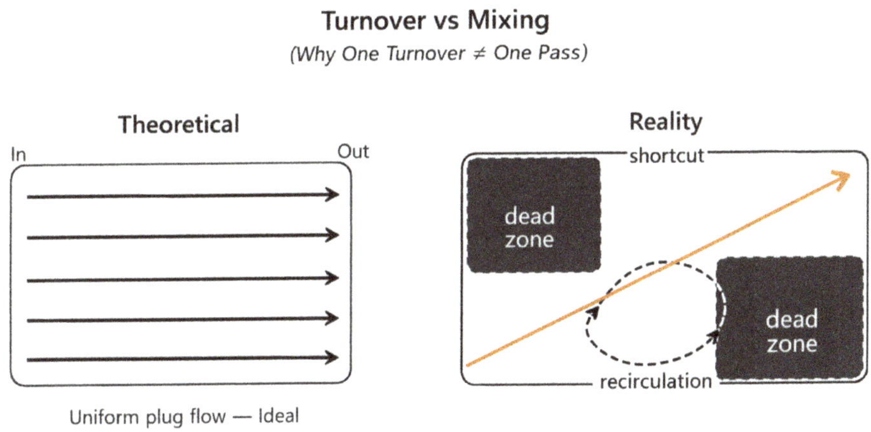

Consider the pool's shape and how water may move within it.

This forms part of the considerations when looking at ORP/pH and any water treatment systems.

Dead Zones and Recirculation Loops

Within a water body, there are often dead zones.

These are usually easy to see.

They are the points where:

- leaf litter accumulates
- algae growth appears
- additional cleaning is always required

What is happening:

These areas have poor circulation.
Water is not being effectively mixed or replaced.

Result:

Chemistry in these areas becomes different from the rest of the pool.

Dead zones can sometimes be improved by:

- redirecting the return eyeball
- adding an automatic cleaner
- changing the circulation pattern

Recirculation loops are a fresh problem.

They are effectively shortcuts.

They occur when water returns directly to the extraction point without mixing through the pool.

Examples include:

- a return jet pointing directly at the skimmer
- surface returns and surface suction working together

Pool Water Circulation System

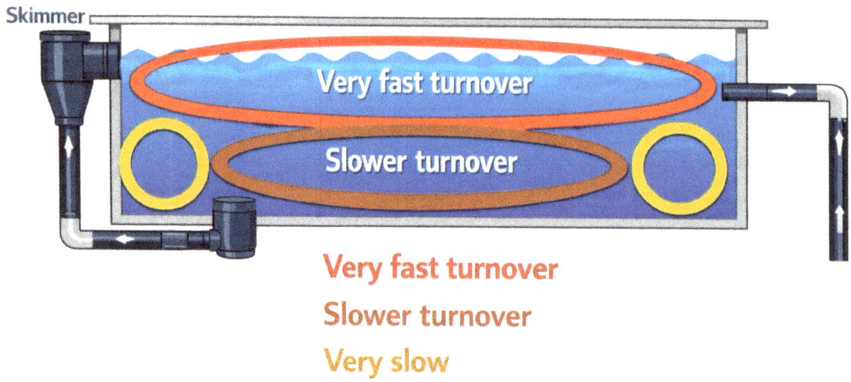

What is happening:

Some water is being turned over repeatedly.
Other parts of the pool are barely being turned over at all.

Result:

The system appears to have good turnover.
In reality, it is uneven.

What this means for ORP:

1. The probe sees water that may not represent the complete pool
2. ORP readings become inconsistent
3. The controller reacts to incomplete information

Laminar versus Turbulent

Laminar (smooth, straight and steady) flow is smooth and undisturbed.

It typically occurs in straight pipe runs of at least five pipe diameters.

Turbulent flow (chaotic, mixed flow) occurs when water changes direction or velocity.

It is commonly found:

- entering and exiting pumps
- around bends and fittings
- at skimmer entry points

What is happening:
Turbulence increases mixing.
But it also introduces instability at the probe.

Laminar vs Turbulent Flow at Probe Locations

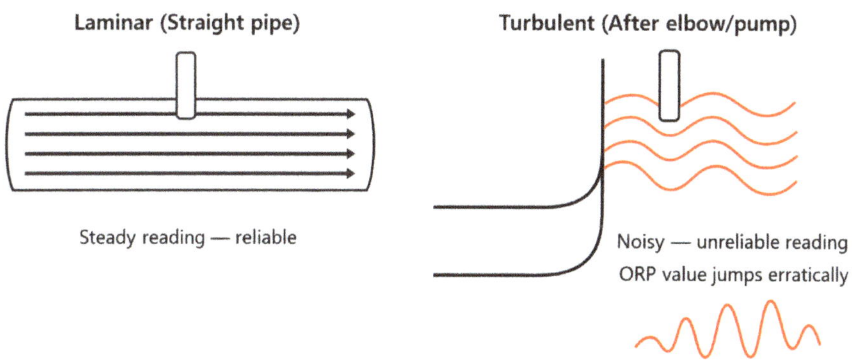

Cavitation

Cavitation (vapour bubbles forming and collapsing in the water) is usually associated with pumps.

Cavitation creates unstable water flow that disturbs probe readings.

It can occur anywhere, but is most common in pumps.

What causes it:

Rapid changes in velocity and direction.

If severe enough, local pressure drops below the vapour pressure of water (about 3.15 kPa at 25°C).

Vapour bubbles form and then collapse violently as pressure recovers.

Result:

- Noise
- Vibration
- Reduced flow
- Long-term equipment damage

Cavitation also introduces additional turbulence.

What this means for ORP:

Turbulent, unstable flow at the probe creates unstable readings.

Rule:

Turbulence creates noise in ORP readings.

pH Changes Induced

Some piping and water treatment equipment can induce pH changes.

What causes this:

Venturis and any form of aeration.

What is happening:

Dissolved CO_2 is driven out of the water in an aerated system.

Result:

- CO_2 decreases
- pH rises
- ORP falls

What this means for ORP:

pH instability feeds directly into ORP instability.

This is covered in more detail in Chapter 16: The CO_2 Connection: Why pH Behaves the Way it Does.

CHAPTER 10

Filtration

Explains why filtration performance must be correct before ORP control can function properly.

Why this matters

If the water is not clean, the ORP reading will not be reliable.

> **Rule:** Dirty water = unreliable ORP.

About

For any chemical dosing system, the filtration system must be working properly and effectively.

What can go wrong:

Problems such as:

- Sizing outside the manufacturer's specifications
- Flow rates higher than intended through the filters
- Different media to specification
- Lack of back-washing or cleaning
- Organic bypassing
- Channelling through the filter bed
- And so on

What is happening:

The water is not being properly filtered.

Particles, organics, and contaminants remain in circulation.

Result:

- Chemistry becomes unstable
- Unwanted reactions occur
- Oxidiser demand increases

What this means for ORP:

- The probe sees a constantly changing chemical environment
- ORP readings become unstable or misleading
- The controller responds to noise instead of real demand

Any filter issues must be addressed before attending to chemical dosing systems.

While this is not a book on filtration, it is worth stating clearly:

ORP/pH control cannot replace adequate turnover combined with effective filtration.

Many times:

Improving circulation and filtration will benefit the water body more than any controller can.

Now that we understand the system surrounding ORP, let us look at what ORP actually is.

CHAPTER 11

How to Read the Chemistry in This Book

Provides a simple guide to the chemical terms and formulas used throughout the book.

While ORP/pH is part of the overall systems used in recreational water treatment, one of those systems is chemistry.

Why this matters:

Some of the behaviour explained in this book cannot be understood without basic chemistry.

For a more complete explanation, chemical formulas are included.

What you need to know:

You do not need to remember or memorise them.

What they are for:

They provide the full mechanism behind what is happening.

They show the chemistry that drives the behaviour you see in the field.

Wherever possible, simple chemistry is used to show each point.

Each equation is also written in words.

This means:

You can follow the idea with no need to interpret the formula.

The chemical terms used are:

Formula	Common Name	What it does in this book
HOCl	Hypochlorous acid	The active, fast-acting sanitising form of chlorine
OCl^-	Hypochlorite ion	The weaker, high-pH form of chlorine
Cl_2	Chlorine gas	The basic form of chlorine dissolves in water to form HOCl
NaOCl	Sodium hypochlorite	Liquid chlorine from a drum
$Ca(OCl)_2$	Calcium hypochlorite	Granular or tablet chlorine
HCl	Hydrochloric acid	Acid most commonly used to lower pH
H_2SO_4	Sulphuric acid	Alternative acid used to lower pH
CO_2	Carbon dioxide	The gas that drives pH behaviour in pool water

H_2CO_3	Carbonic acid	Transitional form of CO_2 dissolved in water; unstable
HCO_3^-	Bicarbonate ion	The main alkalinity buffer in pool water
CO_3^{2-}	Carbonate ion	Higher pH form of alkalinity
$NaHCO_3$	Sodium bicarbonate	Raises total alkalinity
Na_2CO_3	Soda ash / Sodium carbonate	Raises pH
H^+	Hydrogen ion	Determines acidity; directly affects ORP readings
OH^-	Hydroxide ion	Determines alkalinity; produced by salt chlorinators
H_2O	Water	—
CYA / $C_3H_3N_3O_3$	Cyanuric acid	Stabiliser that protects chlorine from UV; reduces active HOCl

NaCl	Sodium chloride / Salt	Used in salt chlorination systems
Cl^-	Chloride ion	End product of chlorine after it has oxidised
NH_3 / NH_4^+	Ammonia / Ammonium	Reacts with chlorine to form chloramines
NH_2Cl	Monochloramine	Primary chloramine; irritant and disinfection byproduct
O_3	Ozone	Strong secondary oxidiser; very short-lived residual
H_2O_2	Hydrogen peroxide	Strong oxidiser used in AOP systems
•OH	Hydroxyl radical	Extremely powerful oxidiser produced by AOP; near-zero lifespan
HSO_5^-	Monopersulphate (MPS)	Non-chlorine shock oxidiser

Cu^{2+}	Copper ion	Found in some algaecides and heat exchangers; poisons ORP probes
Fe^{2+} / Fe^{3+}	Ferrous / Ferric iron	Found in some chemicals and fill water; causes staining and probe fouling
Ag/AgCl	Silver / Silver chloride	Reference electrode material inside ORP and pH probes
KCl	Potassium chloride	Storage and reference solution for probes
E° (V) (standard measure of oxidiser strength)	Standard electrode potential	Measures how strongly a chemical wants to gain electrons
mV	Millivolts	The unit ORP is measured in

This table is here for reference at any point.
There is no expectation that you'll remember it after one reading.

CHAPTER 12

Pool Chemicals

Explains how different chemicals influence ORP and overall water behaviour.

Every chemical we add to a pool affects ORP differently.
Some change it a lot, some barely move it, and some mess it up completely. To understand why, we look at the Standard Electrode Potential — E° (V).

This is just a fancy way of measuring how strongly a chemical wants to grab electrons. And since ORP measures electron activity, this tells us how big an impact it will have.

Simple rule, and one worth remembering:

Higher E° = stronger oxidiser = bigger effect on ORP.

The following shows how common disinfectants compare, roughly, under standard conditions you'd find in a lab.

The table below ranks common disinfectants in order of E° (V).

Oxidiser (Pool-relevant couple)	Formula / Half-reaction (as reduction)	E° (V)	Typical Role / Notes
Fluorine (for reference only)	$F_2 + 2e^- \rightarrow 2F^-$	+2.87	Not used in pools; benchmark for "very strong" oxidisers.

Oxidiser (Pool-relevant couple)	Formula / Half-reaction (as reduction)	E° (V)	Typical Role / Notes
Ozone	$O_3 + 2H^+ + 2e^- \rightarrow O_2 + H_2O$	+2.07	Extraordinarily strong oxidisers; contact-time systems, little residual.
Monopersulphate	$HSO_5^- + 2H^+ + 2e^- \rightarrow HSO_4^- + H_2O$	+1.85	Explains MPS oxidative punch (peroxy bond).
Hydrogen peroxide (acidic)	$H_2O_2 + 2H^+ + 2e^- \rightarrow 2H_2O$	+1.78	Strong oxidiser; no residue; sometimes used for cleanup.
Permanganate (acidic, for comparison in troubleshooting)	$MnO_4^- + 8H^+ + 5e^- \rightarrow Mn^{2+} + 4H_2O$	+1.49	Not standard for pools; occasionally specialty treatments.
Chlorine	$HOCl + H^+ + 2e^- \rightarrow Cl^- + H_2O$	+1.49	The workhorse oxidiser/sanitiser.

Oxidiser (Pool-relevant couple)	Formula / Half-reaction (as reduction)	E° (V)	Typical Role / Notes
Dioxide of chlorine (if used)	$ClO_2 + e^- \rightarrow ClO_2^-$	≈ +0.95 to +1.0 (varies with pH/speciation)	Specialty oxidisers for problem waters; values vary by conditions.
Bromine	$Br_2 + 2e^- \rightarrow 2Br^-$	+1.07	Spa/hot water oxidiser and sanitiser.
Oxygen (acidic)	$O_2 + 4H^+ + 4e^- \rightarrow 2H_2O$	+1.23	Reference for aeration/ORP background; weak vs. Cl_2.
Hypochlorite to chloride (as overall oxidiser strength reference)	$ClO^- + H_2O + 2e^- \rightarrow Cl^- + 2OH^-$	+0.90	Illustrates the lower oxidising power of hypochlorite at high pH.

Special case hydroxyl radical (OH)

Hydroxyl radicals are an extremely strong oxidiser.

Where they come from:

- Advanced oxidation processes (UV/ozone, UV/H_2O_2)
- UV-induced breakdown of chlorine

What is happening
Hydroxyl radicals form briefly, then disappear.
They have a very short lifespan.

Result
They contribute to oxidation in the water.
However:
They do not remain long enough to create a stable ORP signal.

Key point
Strength alone does not determine ORP behaviour.

A chemical can be very strong, but if it is short-lived:

→ it affects demand

→ not the measured ORP value

Important distinction: Strength vs Presence

This is where people get confused.

Ozone is ***stronger*** than chlorine, yes. But it vanishes almost instantly. So even though it pushes ORP way up for a moment, it does not create a stable reading you can control from.

Chlorine is slightly less strong, but it stays in the water for hours. ***That*** is why chlorine is what ORP responds to most of the time.

> **Rule:** *ORP responds to what is present and persistent — not just what is theoretically strongest.*

The CYA fundamental change

Cyanuric acid changes everything. If you take nothing else from this chapter, remember this section.

We add CYA to protect chlorine from sunlight. That is its purpose. But what it actually does is ***clench*** the chlorine molecules tightly.

At normal outdoor levels — 40–50 mg/L CYA, which is pretty much standard in WA — only about 3–10% of your chlorine is actually active HOCl. The rest is bound up, sitting there doing nothing.

What does that mean for ORP?

It drops the reading by 100–150 mV compared to an indoor pool with the same chlorine level.

If you try to use textbook setpoints[1] like 720 mV in an outdoor pool with CYA? You will dose forever, never reach it, and end up with way too much chemical in the water.

> **Rule:** *Setpoints must match **your** water, not generic numbers from a manual.*

ORP is not simply measuring "chlorine". It's measuring the *effective* strength of the sanitiser in *your specific conditions.*

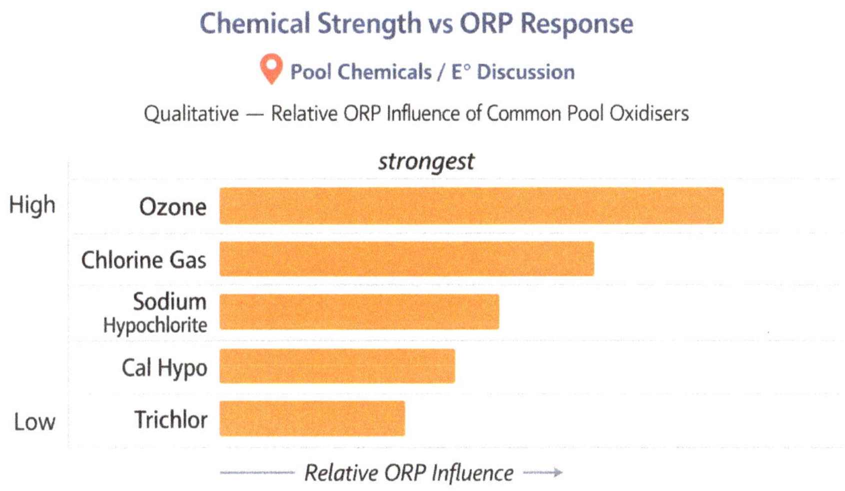

ORP/pH control is unreliable in these settings — do not rely on it alone.

Further reference

Chapter 51 (Appendix: Chemical Potential E° values) lists common pool chemicals and their calculated oxidising strengths.

This information is for reference only.

It is not required for understanding the operation of ORP control.

1 The target value you want the controller to hold.

CHAPTER 13

A Practical View of ORP Control: What is Seen

Describes the common real-world behaviours operators observe in ORP-controlled pools.

Before diving into chemistry, redox theory, or electrochemical models, it is worth describing ORP control as it appears in the real world.

This chapter does not explain why ORP behaves the way it does.

It explains what is observed day to day.

Why this matters:

Most ORP problems are not random.

They follow repeatable patterns.

Experienced operators recognise these patterns quickly:

- A pool that looks stable on Monday but is over-chlorinated by Wednesday
- A spa that behaves perfectly for months and then suddenly cannot be controlled
- A controller that seems “possessed” after a weather change or heavy-use weekend

These behaviours are often labelled “controller failure.”

In reality, they are usually system failures, not instrument failures.

Rule:

The controller responds to signals, not reality.

ORP controllers do not measure safety, cleanliness, or compliance.

They respond to electrical conditions at the probe tip, filtered through software logic.

What this means:

If the signal is unstable → the controller response will be unstable

Understanding this distinction is the first step toward using ORP safely.

When ORP appears to work well:

ORP control behaves best when the pool environment changes slowly and predictably.

Examples:

- Large water volume
- Modest bather load
- Limited aeration
- Stable temperature
- Consistent chemical conditions

What is happening:

The system is stable.

Result:

The controller appears "smart."

When ORP struggles:

ORP control struggles when the environment changes quickly.

Examples:

- Spas
- Toddler pools
- Spray pads
- Highly aerated water bodies

What is happening:

The water is changing faster than the control loop can respond.

Result:

- Oscillating readings
- Overshooting chlorine
- Frequent manual intervention

This is not because the controller is broken.

It is because the system is unstable.

Common operational mistake:

Stable numbers = stable water

This is not true.

An ORP reading can remain constant because:

- The controller is averaging the noise
- The system is lagging actual changes
- Multiple errors are cancelling each other out

Result:

False confidence

Therefore, every standard requires manual testing, regardless of automation.

Another common pattern:

pH problems appearing as chlorine problems

What is happening:

pH drifts → ORP drifts

Operator response:

Adjust chlorine setpoint

Actual cause:

- pH instability
- CO_2 loss
- Aeration
- Heating
- Bather activity

Result:

A cycle of correction followed by worse behaviour

Important point:

ORP controllers usually do exactly what they are told to do.

When a controller feeds an entire drum of chemical into a pool, it is rarely because it has failed.

It is responding correctly to the signal it is receiving.

The real problem:

The signal is wrong.

This may be caused by:

- Probe fouling
- Air
- Temperature change
- Chemistry, the probe cannot distinguish

Operator mindset:

Treat ORP as an assistant, not an authority.

It can help maintain stable conditions.

It cannot stabilise an unstable system.

The water must be stabilised first through:

- Hydraulics
- Filtration
- Sensible chemistry

Only then should automation be trusted.

This chapter shuns equations, redox couples, and electrochemical theory.

Purpose:

To anchor the reader in operational reality.

Once these patterns are familiar, the deeper explanations that follow will feel like answers to real problems, not abstract science.

If this behaviour already feels familiar, you are ready to move on.

The next chapters will explain why these patterns occur and why ORP behaves the way it does beneath the surface.

CHAPTER 14

ORP What Is It?

Defines ORP in practical terms and explains what the probe is measuring.

ORP stands for Oxidation (loss of electrons) Reduction (gain of electrons) Potential.

ORP is also called REDOX (reduction-oxidation potential) (oxidation and reduction happening together).

What ORP is NOT

ORP is not a chlorine measurement.

It does not measure the disinfectant directly.

That distinction matters, because if you treat it like a chlorine meter, it will give you the wrong answer every time.

What ORP is measuring

ORP is a measurement of the electron activity in the water.

That activity represents the overall oxidation condition of the system at the probe location.

Think of ORP as the water's electrical mood, reflecting how ready it is to react chemically.

Why this matters

ORP does not respond to one thing.

It responds to multiple systems acting at the same time:

- The chlorine system
- The carbonate system

- Hydraulic mixing
- Probe conditions

Result:

ORP reflects the combined behaviour of the system.

It does not isolate individual causes.

Basic Redox Concept

In its simplest form:

- An oxidiser accepts electrons
- A reducer donates electrons

A simple representation is:

$$A + B \rightarrow AB$$

For example:

$$2H2 + O2 \rightarrow 2H2O$$

This example is simplified.

It shows how substances combine through electron movement.

Important clarification

This type of reaction does NOT describe what is happening in the pool system as a whole.

It describes what is happening at the probe:

- where electrons are being exchanged

Water behaviour

Water is not chemically static.

It is always in equilibrium.

For example:

$$H_2O \rightleftharpoons H^+ + OH^-$$

This is not a redox reaction; no electrons are transferred.

But it matters because:

This equilibrium controls:

- pH
- conductivity
- electrochemical behaviour

Important note:

At a molecular level, these reactions happen extremely fast.

Faster than our instruments can directly observe.

Result:

pH appears stable — but it is actually dynamic.

A clear redox example

Rusting of iron:

$$4Fe + 3O_2 + 6H_2O \rightarrow 4Fe(OH)_3$$

4 by Iron + 3 by Oxygen + 6 by Water =>

4 Ferric Hydroxide (Rusting)

What is happening:

- Iron loses electrons (oxidation)
- Oxygen gains electrons (reduction)

The oxidising agent (oxygen) gains electrons.

The reducing agent (iron) donates electrons.

What ORP measures

ORP measures the movement of electrons.

Electrons also flow in electrical circuits.

In electrical systems, this flow is measured in volts.

In ORP:

Electron movement occurs within chemical reactions.

These reactions do not produce usable electricity.

To measure them, we provide a path using the probe.

How the probe works

The ORP sensor has two electrodes:

- A platinum or gold measuring electrode (the sensing surface)
- A reference electrode (silver or silver chloride)

What is happening at the probe:

- Oxidisers try to remove electrons from the measuring electrode
- Reducers try to add electrons

Result:

This creates a tiny voltage.

Measurement

This voltage is measured in millivolts (mV).

Final interpretation

ORP is not the chlorine level.

It is not a direct safety measure.

It is a measurement of the oxidation condition of the circulating system at the probe location.

In simple terms:

ORP is the electrical mood of the water.

ORP Probe: How the Electrodes Actually Work

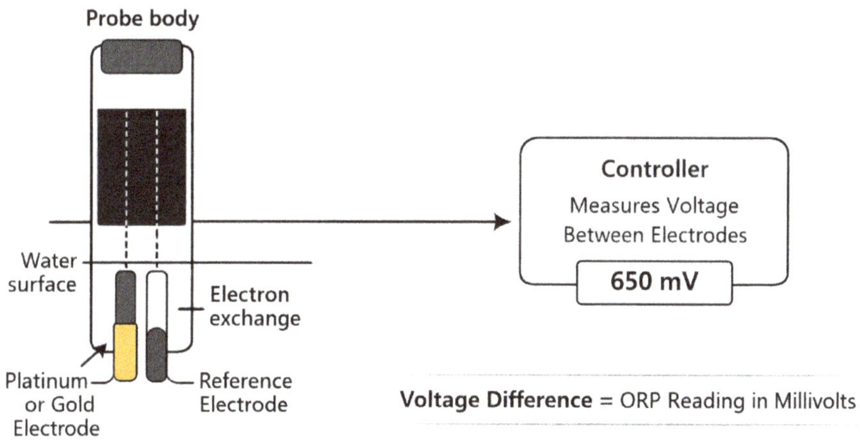

CHAPTER 15

Elevate ORP

Demonstrates how ORP can change because of factors other than chlorine.

Previously, we established that ORP is a measure of electron flow.

In a water body context, we can use this measurement to show the likely sanitisation level of the water.

Well, that is the theory, at least.

Consider this example:

- Pool 50,000 litres
- No filtration running
- pH 7.8
- Active algae growth
- No chlorine

What happens during the day

Algae produce oxygen through photosynthesis.

Effect 1:

Dissolved oxygen increases

- ORP increases

(Oxygen is an oxidiser)

Effect 2:

CO_2 is consumed

- → pH increases

Combined result:

- Oxygen pushes ORP up
- Rising pH pushes ORP down

The oxygen effect dominates.

Result: ORP is relatively high during daylight (often 500–600 mV)

What happens at night

Photosynthesis stops.
Algae respire.

Effect 1:

Oxygen is consumed

- → ORP decreases

Effect 2:

CO_2 is released

- → pH decreases

Combined result:

- Lower pH pushes ORP up
- Loss of oxygen pushes ORP down

Losing oxygen dominates.

Result: ORP falls overnight

Key point:

ORP is responding to multiple processes at once.

Note:

In a chlorinated pool, a drop in pH will usually increase ORP.

Biofilm

Biofilm (a layer of microorganisms on surfaces) behaves similarly.

It contains algae and other organisms that:

- Produce oxygen during the day
- Consume oxygen at night

Result: ORP and pH can vary by time of day

Practical demonstration: Bucket test

This example shows the same behaviour in a simple system.

Items needed:

- Clean bucket
- Pocket ORP probe
- pH test kit or tester
- 300 mg salt
- 1 litre deionised water
- 10 ml of vinegar
- Paint stirrer (for a power drill)
- Power drill

Method:

Pure water

Add 1 litre of deionised water.
Try to measure ORP.

What happens: No stable reading

Why:

- Pure water has almost no ions
- ORP requires conductivity
- The probe cannot form a stable circuit

Result:

ORP in very low TDS water is meaningless

Add salt

Add 300 mg of salt and mix.
Measure ORP.

What happens: Reading stabilises (~+80 to +120 mV)

Why:

- Conductivity is now sufficient
- The probe can form a circuit
- No oxidisers added

Add acid

Add 10 ml of vinegar.
Measure pH and ORP.

What happens:

- pH decreases
- ORP increases (often 50–150 mV)

Why:

Lower pH increases the oxidising strength of dissolved oxygen

Key point:
No chlorine was added

Step 4

Same water
no chlorine added

ORP: HIGH

Add aeration

Stir vigorously with a drill.

What happens:

- Oxygen increases
- CO_2 leaves
- pH increases
- ORP increases further (20–80 mV)

Why:

- Oxygen is an oxidiser → raises ORP
- CO_2 loss changes pH → affects ORP

Final observation

We changed the ORP without adding chlorine.

Cause of ORP movement:

- Oxygen (from the air)
- pH change
- CO_2 loss

Key conclusion

ORP does not depend only on chlorine.

It responds to:

- Oxygen
- pH
- CO_2
- biological activity

Connection back to real pools

This is exactly what happens with algae:

- Day: oxygen production → ORP rises
- Night: oxygen consumption → ORP falls

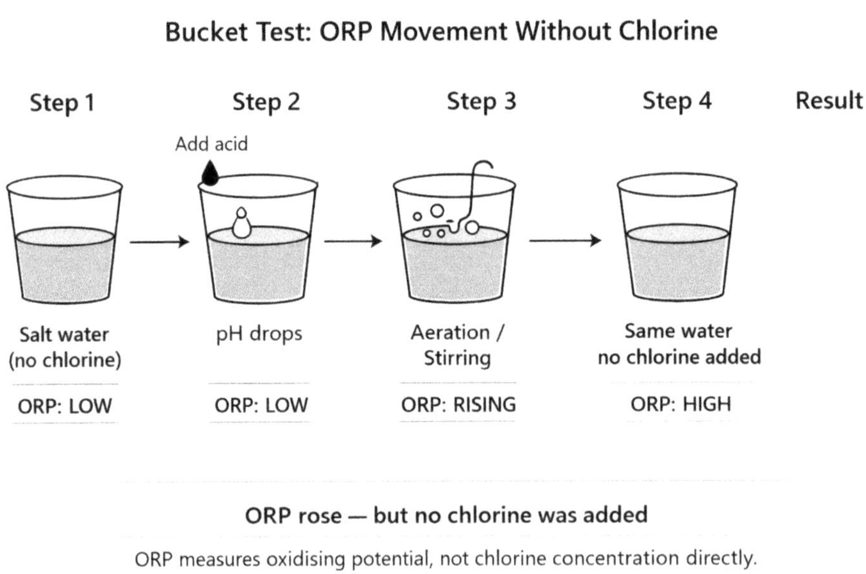

CHAPTER 16

The CO_2 Connection: Why pH Behaves the Way It Does

Explains how CO_2 controls pH behaviour and why this drives ORP changes.

Why this matters:

pH movement is actually CO_2 movement — and ORP follows pH.

Rule:

pH follows CO_2

When water is disturbed:

CO_2 leaves → pH rises → ORP falls

About

As I wrote this book, it became increasingly clear that an understanding of CO_2 chemistry is required to understand ORP/pH control.

This chapter covers only what is needed.

If it feels like a recap, move on quickly.

If it is new, treat it as essential.

CO_2 loss means carbon dioxide gas escapes from water into the air, raising the pH.

What happens in practice

Water loses CO_2 when disturbed.

Result:

CO_2 loss → pH rises → ORP falls

This is the dominant behaviour in most pools.

How it works

All water bodies using acid are using CO_2 for pH control.

A simplified reaction:

HCl + NaHCO3 → NaCl + H2CO3 → NaCl + H2O + CO2(aq)

Hydrochloric acid + Sodium bicarbonate => Sodium Chloride + Carbonic Acid => Salt + Water + Carbon dioxide (aqueous)

What is happening:

Carbonic acid forms temporarily.

It is unstable.

It converts into CO_2.

CO_2 then leaves the water through the surface.

Gas exchange

The water surface controls how easily CO_2 escapes.

Gaseous interchange layer:

This is how quickly gases move between water and air.

It controls:

- CO_2 leaving (and entering)
- oxygen entering

What increases CO_2 loss:

Anything that disturbs the surface:

- Blowers
- Turbulence
- Bather activity
- Fountains
- Weirs
- Surface movement of any kind

Result of surface disturbance

More disturbance → faster CO_2 loss

Chain:

CO_2 leaves → pH rises → ORP falls

System response

As the pH rises:

More acid is required.

What the acid does:

Acid converts alkalinity into CO_2.

This reduces total alkalinity (buffering capacity).

Practical consequence

The system becomes:

- more reactive
- less stable

Operator reality

There is only one practical way to manage this:

Reduce avoidable aeration where possible.

(If that means controlling features or usage, then do so.)

Impact on ORP

Aeration-driven pH rise directly affects ORP readings.

Result:

Unstable pH → unstable ORP

Key clarification

Yes, adding acid directly adds hydrogen ions.

But in pool water, the dominant effect is:

Acid converts alkalinity into dissolved CO_2.

CO_2 concentration controls pH.

Plant room interpretation

pH does not drift randomly.

It follows CO_2.

When the water is disturbed:

CO_2 leaves → pH rises → ORP falls

This happens continuously.

It happens faster in:

- warm conditions
- high activity periods

Operational meaning

The controller is not malfunctioning when the ORP drops during use.
It is responding to real chemistry.

What to do

Stabilise total alkalinity.

This slows pH movement.

Avoid reacting too quickly during high activity.

The system will partially stabilise once the disturbance reduces.

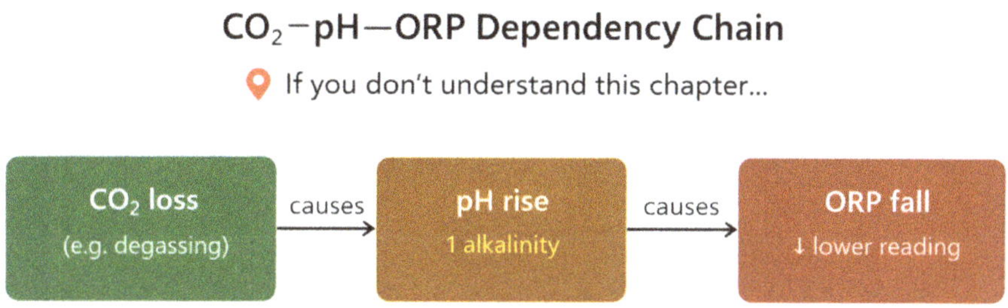

CO_2 is the hidden driver — control CO_2 and you help stabilise ORP

Reference values (example)

TA (mg/L as $CaCO_3$)	pH	CO_2 (mg/L, aqueous)
200	7.5	20.05
180	7.5	18.04
160	7.5	16.04
140	7.5	14.03
120	7.5	12.03
100	7.5	10.02
80	7.5	8.02
60	7.5	6.01

CO_2 controls pH.

pH controls ORP.

Disturbance controls CO_2.

For deeper detail
For a more in-depth explanation, see my book Carbon Dioxide for Pool pH Control.

CHAPTER 17

Temperature and ORP – What Really Matters

Explains how temperature affects ORP both directly and through water chemistry.

Temperature moves ORP a little, and it also changes how pH behaves.

Why this matters

Temperature changes ORP in two ways.

Only one of them matters in practice.

Rule:

More heat → more CO_2 loss → higher pH → lower ORP

What you see in real pools

Temperature changes usually appear as a slow drift in ORP readings.

Typically:

A seasonal shift of 10°C

- → about 20–30 mV ORP change

This is normal.

Two ways temperature affects ORP

1. Probe physics (slight effect)
   Electrochemical measurements shift slightly with temperature.

What is happening:

The probe itself changes its response with temperature.

Typical scale:

~2–3 mV per °C

Result:

Small, predictable drift

Important:

This is not a fault.

It is how the measurement works.

2. Water chemistry (dominant effect)
   This affects control.

What is happening:

Warmer water loses CO_2 faster.

Chain:

Temperature ↑

→ CO_2 loss ↑

→ pH rises

→ ORP falls

Result:

ORP drift that looks like a chlorine problem

Key concept to remember

- ORP follows pH
- pH follows CO_2
- CO_2 follows hydraulics, aeration, and temperature

Where this shows up

- Outdoor seasonal pools
  Spring → summer
  → ORP drifts without dosing changes

- Wind-exposed or shallow pools
  → faster CO_2 loss
  → unstable ORP

- Heated indoor pools
  Stable air ≠ stable pH
  → heating still drives CO_2 loss
  → ORP changes during the day

- Spas
  Heat + aeration
  → rapid CO_2 loss
  → rapid pH rise
  → strong ORP swings

What this means operationally

The controller is reacting to temperature-driven chemical changes.

It is not malfunctioning.

What to do BEFORE adjusting anything

1. Confirm pH is stable
   If the pH moves, ORP will move

2. Check for added aeration
   → features, wind, covers, programs

3. Check recent temperature changes
   → weather, heating, seasonal shifts

Rule:

If the pH is not stable, ORP cannot be stable.

If ORP is off target but pH is stable

Treat it as temperature drift

1. Adjust slowly
   → ±5 mV at a time
2. Wait for system response
   at least one turnover (ideally 24 hours)
3. Keep a simple log
   → temperature, pH, chlorine, ORP, usage

If the pH is not stable

1. Fix pH stability first
   → control CO_2
   → reduce aeration
   → check alkalinity and acid feed
2. Re-check ORP

Result:

The ORP problem often disappears

Rules of thumb

- Seasonal drift of tens of mV is normal
- Do not chase ORP when pH is moving → this causes chlorine overshoot
- Heated and aerated systems need wider control tolerances

How controllers handle temperature

Most controllers compensate pH for temperature.

They do not fully compensate ORP.

Why:

The ORP response depends on the actual chemistry in the water. Not just the probe.

Practical meaning

Small ORP drift from temperature is normal.

You manage it operationally:

- keep the pH stable
- adjust setpoints slowly

CHAPTER 18

ORP/pH Relationship

Shows how changes in pH directly move ORP values in predictable ways.

What came first — the chicken or the egg?

pH is a lot like that tale.

Which is more important: the sanitiser level or the pH?

For ORP, the answer is pH.

Clearly.

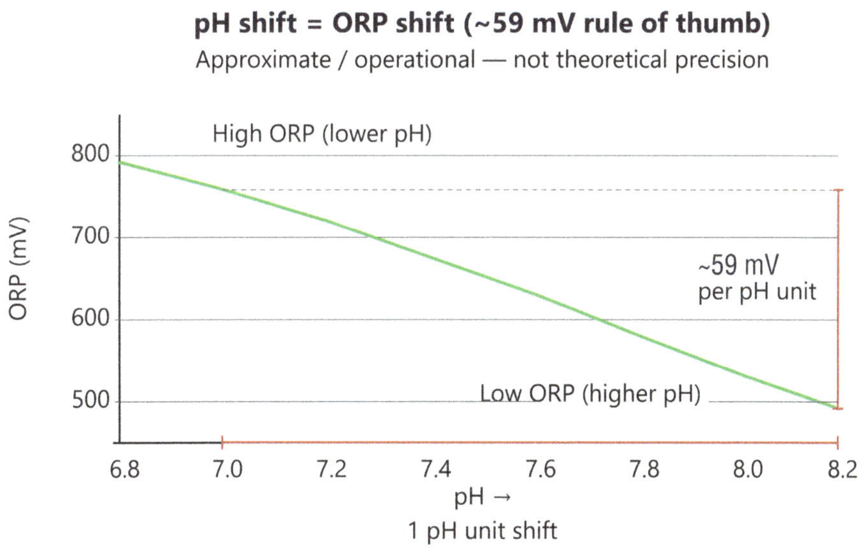

Core idea

ORP moves with pH.

You already saw this in the bucket test.

When we added acid, the ORP increased — even with no chlorine.

Rule:

pH movement is ORP movement.

What the chemistry says

Each point of pH change shifts ORP by approximately 59 mV:

+59 mV when pH goes down

–59 mV when the pH goes up

This comes from the Nernst equation (simplified relationship between pH and ORP under ideal conditions). (in a simple chlorine-free system at 25 °C, theoretical Nernst response).

The Nernst Equation (simplified)

$E = E^\circ - (0.0592/n) \times \log [products/reactants]$

Where:

- **E** = *measured ORP (mV)*
- **E°** = *standard potential of the reaction*
- *n = number of electrons transferred*
- **0.0592** = *a temperature constant (at 25°C)*

What this means in simple terms

As pH drops:

→ Oxidising strength increases

→ ORP increases

As pH rises:

→ Oxidising strength decreases

→ ORP decreases

This relationship is logarithmic, not linear.

Result:

Small pH changes can create large ORP shifts.

Why this happens

Lower pH increases the oxidising power of both oxygen and chlorine.

Key point:

The probe responds to oxidising strength.
Not just the chlorine quantity.

Plant room reality

pH movement is ORP movement.

If the pH rises by 0.5,

→ ORP drops by ~25–30 mV

This happens even if chlorine has not changed.

What the controller does:

It sees the ORP drop.

It doses chlorine.

Problem:

It cannot tell that pH caused the change.

What to do

Before adjusting anything:

Check pH.

If pH has moved:

That is the cause.

Examples

Example 1 — Drop in pH

7.8 → 7.3

Change = 0.5

→ ORP rises ≈ 30 mV

Example 2 — Aeration

Oxygen increases

→ ORP rises (≈ 20–50 mV)

(Bucket test result)

Example 3: Add acid

pH decreases

→ ORP increases

Even with no chlorine

Why this matters

ORP is strongly affected by pH.

If the pH is unstable:

→ ORP will be unstable

pH moves naturally due to:

- Aeration
- CO_2 loss
- Bather load
- Temperature
- Chlorination

System behaviour

ORP control depends on pH stability.

pH stability allows ORP to behave predictably.

pH does not determine ORP, but it controls how ORP behaves.

Final rule

Unstable pH = unstable ORP = unstable control.

CHAPTER 19

Ideal Pool

Describes the theoretical conditions where ORP control behaves consistently.

What ORP needs to work well

For ORP/pH control to behave predictably, the system needs to be stable.

An ideal pool would be:

- Large enough for chemicals to stabilise (50,000 litres minimum)
- Held at a stable temperature, ideally
- Free of aeration, features, and waterfalls
- Supplied with clean, filtered fill water (low/no metals)
- Not using algaecides or clarifiers
- Free from bather load
- Free from cyanuric acid
- Indoors and sheltered from the weather
- Running at a low chlorine level
- Supported by a secondary sanitiser (ozone or UV)
- Free of stray induced currents

Result:

ORP and pH remain stable.

Reality
Real pools are not like this.

Rule:

ORP works best in calm, predictable systems.

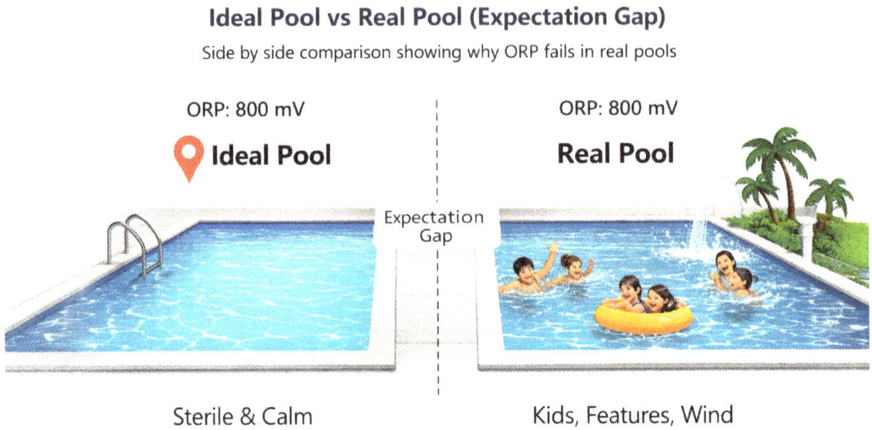

Where ORP/pH should not be relied upon

At the other end of the scale, we have the systems that are practically the opposite of that ideal list. You have seen them; I have seen them; every operator has been caught out by them.

Systems with constant disturbance

These are effectively the opposite of the ideal pool.

Examples

- Spray pads
- Spas
- Wave pools
- Lazy rivers
- Highly aerated leisure pools

What they have in common

- Small or variable water volume
- High aeration
- Constant disturbance
- Rapid changes in load

What is happening

Every time you splash, every time a jet runs, every time the wind blows across the surface — CO_2 leaves the water. And remember our rules?

pH follows CO_2. CO_2 follows disturbance.

So, disturbance → CO_2 leaves → pH rises → ORP falls.
It happens fast — sometimes in minutes.

Result

In these systems, ORP control becomes a guessing game. You get:

- Unstable readings
- Overdosing
- Underdosing

And you end up visiting every second day just to keep things safe.

Important clarification

This does not mean ORP "does not work".

ORP/pH control is unreliable in these settings — do not rely on it alone.

It means:

ORP is conditional.
Some systems simply cannot meet those conditions.
ORP only works when the conditions allow it. If the system itself is unstable, the controller cannot fix it. It will only react to the chaos.

Example: Spray pads

Take spray pads, for example.

- tiny volume.
- water shooting everywhere.
- constant splashing.

Result:

I have seen chlorine levels hit 30 mg/L or more because the controller was trying to chase a signal that was changing every few seconds. That is not a failure of the equipment — it is a mismatch between the tool and the job.

Real-world consequences:

Poorly applied ORP control has caused chlorine spikes (>30 mg/L) and injuries.

Example: Spa pools

Spas amplify every instability factor:

What is happening

- Aeration (jets/blowers) → rapid CO_2 loss
- Rapid pH rise
- Highly variable bathing load
- Small volume
- Fast turnover (incomplete reaction time)

Constantly shifting chemistry.

Effect on ORP

Unstable signal → unstable control

Initial observation

System installed → readings stable → everything appears correct

Case study: Illusion of control

This is one of the most common traps.

Initial observation

You install the system, everything looks great, readings are flat and steady, and you think: "Perfect, this one is set and forget."

Delayed reality

Then a busy weekend hits, or the weather warms up, or usage increases permanently.

What happens

- Aeration increases
- pH rises
- ORP drops
- Controller doses hard

What confuses operators

Days later:

A few days later, things calm down, the numbers look normal again, and you think: "Oh, it fixed itself."

Why:

It did not fix itself. You just got lucky. Next time it will be worse. The system has smoothed itself out.

Then the system changes

Usage increases permanently

(new tenant, seasonal change, more activity)

Result

- pH instability increases
- Chemical demand increases
- ORP swings

Outcome

- Overdosing
- Underdosing
- System becomes erratic

Often:

The spa is abandoned.

Key lesson

This is not controller failure.

Here is the hard truth I have learned over forty years, and I will say it again because it is that important:

> *Stable numbers do NOT mean stable water.*

A flat reading for hours can mean the controller is averaging out the noise, or the system is lagging what is actually happening, or two distinct problems are cancelling each other out.

This is exactly why every standard, every code, every sensible operator still does manual testing — no matter how good the automation looks.

The right way to think about it

Standing here next to the plant room, this is how I explain it to everyone:

Treat ORP as your ***assistant,*** not your manager.

It will help you maintain stability once you have created it. It will never create stability out of chaos.

Get the hydraulics right, get the filtration right, get the chemistry balanced and stable — ***then*** turn the controller on and let it do what it is good at.

If you hand it an unstable mess? It will make it worse, guaranteed.

CHAPTER 20

Where Operational Rules Finally Break Down

Explains why simple rules stop working when real chemistry and system interactions take over.

What this chapter does:

Everything I have told you so far works perfectly — until it does not. And this is the chapter where we explain why the simple rules stop working, why sometimes ORP behaves as if it has a mind of its own.

A lot of what I will explain here comes directly from the work of Richard A. Falk, whose deep dive into chlorine chemistry cleared up so much confusion in this industry. I want to acknowledge that upfront because without his work, we would still be guessing.

You do not need to memorise this.

But if you have ever seen:

- ORP rising with no chlorine change
- ORP falling while dosing increases
- ORP not matching test results

This explains why.

Core idea

ORP does not respond to chlorine alone.
It responds to how chlorine behaves in the system.

Chlorine in water (speciation)

When you add chlorine to water, it splits into two forms — always, every time:

- Hypochlorous acid (HOCl)
- Hypochlorite ion (OCl^-)

What controls this?

- pH.

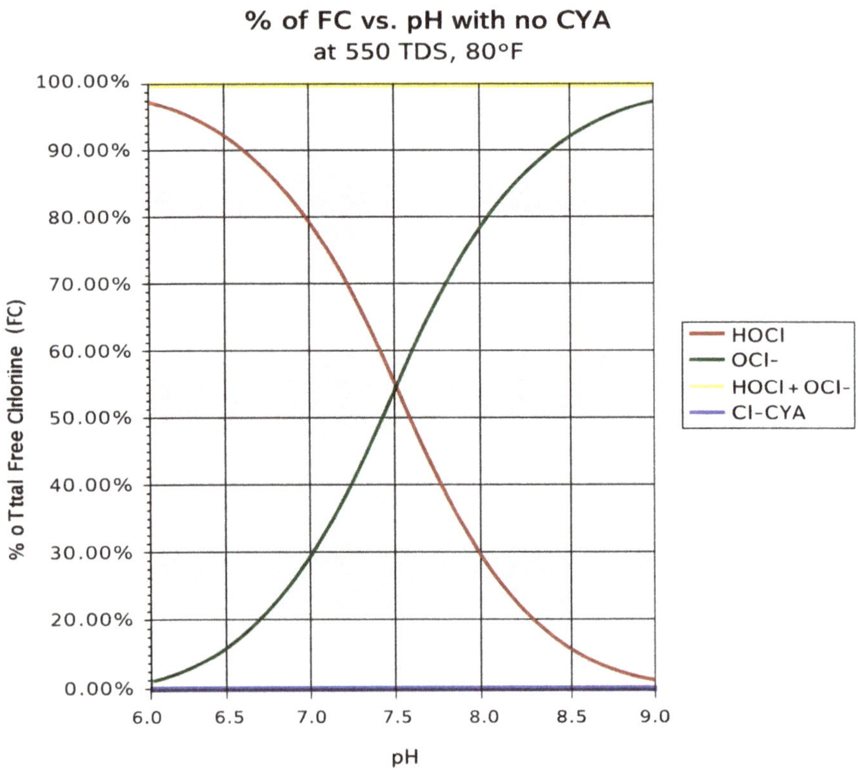

At about pH 7.5:

→ roughly 50% HOCl

→ 50% OCl^-

- At pH 7.2 → ~65% is active HOCl
- At pH 7.8 → only ~25% is active HOCl

As pH rises:

→ HOCl decreases
→ OCl^- increases

This means the same amount of chlorine added can have completely different effective strengths depending on pH, resulting in a completely different ORP reading.

Why this matters

HOCl and OCl^- behave differently.

- HOCl = fast, effective disinfectant
- OCl^- = slower, less reactive

First key point

pH changes the form of chlorine.

What the probe actually sees

The ORP probe does not "see chlorine."
It sees an electrochemical reaction on its surface.

The reaction:

$$HOCl + H^+ + 2e^- \rightleftharpoons Cl^- + H_2O$$

Hypochlorous acid + Hydrogen ion + 2 electrons ⇌ Chloride + Water

In review I don't think:

What this means

Hydrogen ions (pH) are part of the reaction itself.

Is sufficiently explaining this point.

Hydrogen ions (pH) are part of the reaction itself. pH has a large impact on ORP at the reaction level.

Second key point

pH affects ORP in two ways at once:

Changes chlorine speciation (HOCl vs. OCl^-)

Directly affects the electrochemical reaction

Result

Even if chlorine stays constant:

→ ORP will still move with pH

The probe cannot separate these effects.

It reports the combined signals.

Why theory does not match reality

In theory:

ORP should follow a predictable relationship (Nernst behaviour).

In real pools:

It does not.

Why

Pool water is not a clean system.

It contains:

- multiple oxidisers
- multiple reducers
- surfaces
- biological activity

The probe responds to all of these at once.

Result

- ORP does not follow textbook behaviour
- No correction factor will fully fix this

Third key point

ORP is not a calculation.

It is an environmental signal.

What does this mean in the plant room?

The chlorine you measure ≠ the chlorine ORP responds to

Test kit

Measures:

→ total free chlorine (HOCl + OCl^-)

ORP probe

Responds mainly to:

→ HOCl (fast-reacting species)

Example

At pH 7.2 → ~65% HOCl

At pH 7.8 → ~25% HOCl

Same chlorine reading.

Different ORP.

Result

Two pools can have:

→ identical chlorine

→ very different ORP

And both are correct.

Why ORP seems to "follow HOCl"

This is not because HOCl is a stronger oxidiser.

It is because:

HOCl reacts much faster.

Key distinction

- Thermodynamics (potential) = how strong the chemical is (same for both forms)
- Kinetics = how fast it reacts (massive difference between them)

ORP measures reaction speed, not just strength.

ORP sees HOCl clearly because it reacts fast. It barely sees OCl^- at all because it reacts too slowly.

Fourth key point

ORP reflects reaction speed, not just chemical strength.

What ORP is really measuring

Not chlorine quantity.

It reflects:

→ effective oxidising activity of the water

Simple comparison

- DPD = quantity
- ORP = effectiveness

Cyanuric acid (CYA) makes it worse

CYA changes everything and makes it much worse.

We already mentioned CYA binds up chlorine. Now add this:

When only 3–10% is active, the relationship between chlorine and ORP becomes very unstable.

- Minor changes in chemistry = huge changes in reading
- Readings wander all over the place
- Control becomes erratic
- You get massive overshoots or underdosing

Result

- Readings wander
- Control becomes unstable
- Overshooting occurs
- Undershooting occurs

Why

ORP is not linear.

A small ORP change can mean:

- large chlorine change

OR

- small chlorine change

This is the biggest reason outdoor pools are harder to control than indoor ones. Generic setpoints can lose meaning here.

Effect of increasing CYA

More CYA

→ less free HOCl

→ lower ORP

Fifth key point

CYA reduces the active chlorine seen by ORP.

What this does to control

Because only a small fraction is active:

→ ORP becomes a lot more variable

Figure: HOCl vs pH with CYA (after Falk, 2019).

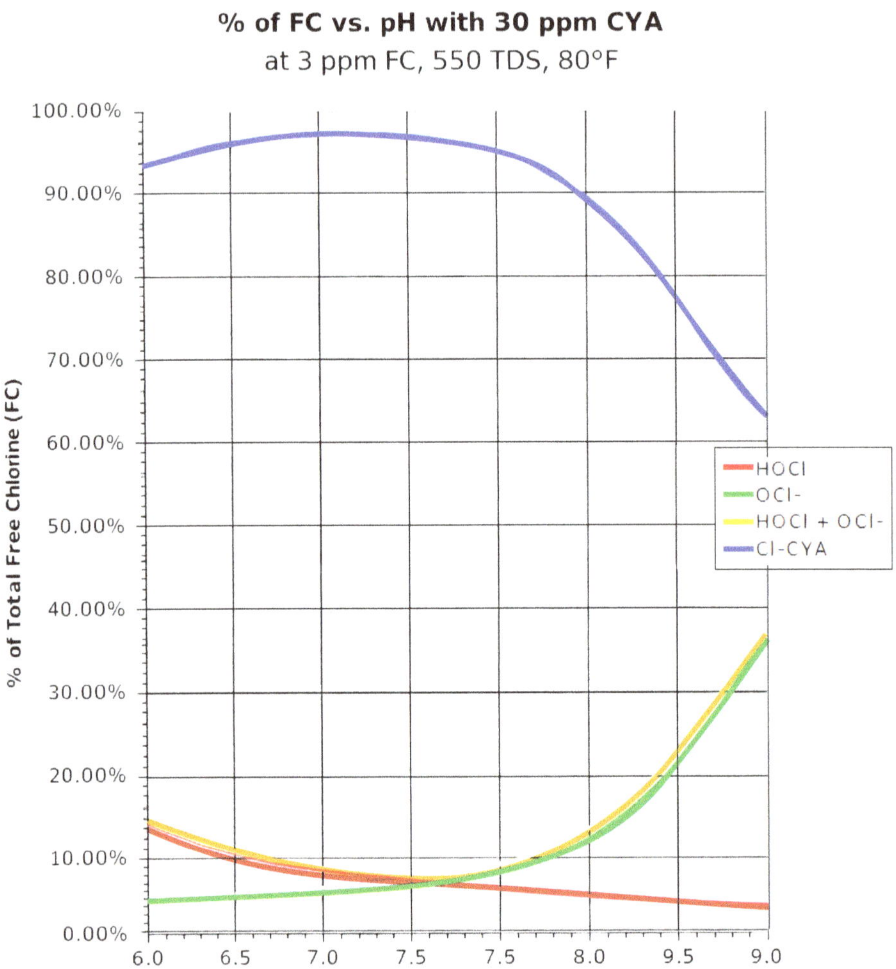

Controller problem

The controller cannot tell the difference.

Definitions

Overshooting:

Adding too much chemical (passing the target)

Undershooting:

Not adding enough (stopping short of the target)

Plant room reality

If your outdoor pool runs:

CYA = 40–50 mg/L (common in WA)

Then:

ORP will run 100–150 mV lower than an indoor pool

- Same chlorine
- Same pH

What this means

Setpoints must match the water.

Not a generic value.

Rule

Use references based on your conditions, not textbook numbers.

Final concept

ORP is not a calculation. It is an environmental signal.

It is affected by:

- pH
- CYA
- chlorine form
- reaction speed
- system conditions

No formula fixes this. No magic setting fixes this.
Only understanding why it happens lets you manage it.

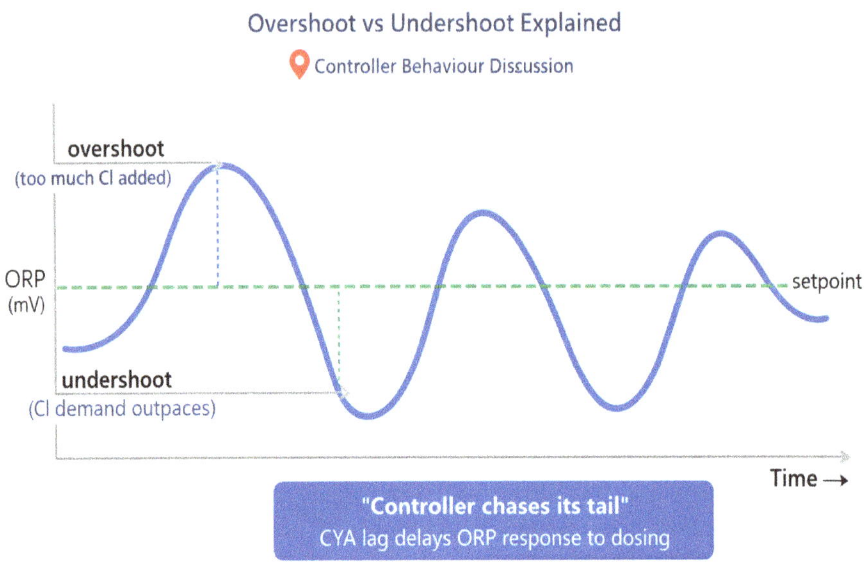

Final takeaway

ORP is not measuring chlorine.

It measures how the system behaves.

CHAPTER 21

Stabilise pH to Stabilise ORP

Explains why a stable pH is required before ORP control can behave predictably.

Stabilise pH, and ORP will largely take care of itself.

Core idea

pH is not a fixed point.
It is constantly moving.

Why it moves

The rate of pH change depends on how quickly CO_2 leaves the water.

Because pH always follows CO_2, and CO_2 always follows disturbance. That chain never breaks.

Every time you splash, jet, heat, or blow air into water, you help CO_2 escape. It is simply physics.

What this means

More disturbance

→ more CO_2 loss
→ faster pH rise

Example: high-disturbance system

Consider a toddler pool:

- Multiple features
- Bucket dumps
- Shallow water
- Wind exposure
- Heavy use
- Constant movement

Result:

Continuous aeration

→ continuous pH movement

Key outcome

pH instability → ORP instability

It is that simple, and that constant.

Where this becomes a problem

There are some water bodies where you can fight this forever.

Typical examples

- River rapids features
- Wave pools
- Rooftop pools
- Exposed coastal pools
- Shallow leisure pools
- Spa pools
- Lazy rivers
- Spray pads
- Other highly turbulent systems

What they have in common

- High aeration
- High disturbance
- Rapid CO_2 loss
- Constant pH movement

Result

Unstable pH

→ unstable ORP

→ unstable control

Human activity matters

Bathers significantly increase aeration.

Example:

A lap pool during medley races

- Butterfly
- Freestyle
- Backstroke

All increase turbulence.

Result:

pH changes during use.

System decision

If a system cannot maintain a stable pH:

ORP/pH control will struggle.

In these cases, alternative technologies like amperometric sensors are often a better choice. If the budget or setup does not allow that? Be honest with the client — ORP is likely not suitable here.

That is not a failure. That is professional responsibility.

Professional responsibility

Understanding limits is not avoidance.

It is the first professional skill.

Expectation management

Many problems are not technical failures.

They are expectation failures.

Not because:

- The product does not work
- The installation is poor

But because:

the expected outcome is unrealistic.

Common expectation

"The system will fully control chlorine and pH at all times with no intervention."

This is not achievable.

Reality

An ORP/pH controller:

- helps maintain conditions
- but requires periodic intervention

Better explanation to clients

"This device will control chlorine and pH under most conditions.

However, it will require occasional correction."

pH myths in the industry

Much industry knowledge is passed on verbally.

Not all of it has been verified.

Example: sunscreen and pH

Common belief:

Sunscreen and cosmetics raise the pH.

Why this seems reasonable

- "Milk and honey" style branding
- Association of dairy products
- Milk is alkaline

Actual observation

Published data shows:

- Many sunscreens are acidic
- Many creams and foundations are also acidic

Conclusion

This belief is folklore.

Passed on through experience, not measurement.

CHAPTER 22

Chemical Dosing with an ORP Controller Installed

Sets out a way to safely add chemicals without interfering with the controller.

Reality of operation

In an ideal world, the ORP/pH controller maintains the pool perfectly.

No manual correction is required.

Reality is different.

On a service visit, you will probably need:

- sodium bicarbonate
- hydrochloric acid
- calcium chloride
- chlorine
- cyanuric acid

First rule

Turn off the ORP/pH controller before adding any chemicals.

We do not want the controller to react to what we are doing.

If possible:

Isolate the probes.

Concentrated chemicals can permanently damage them.

How to add chemicals safely

These are general steps.

Always follow the manufacturer's instructions and the normal treatment order.

Domestic pools

1. Make sure the pool is not in use
2. Ensure equipment is running and filters are clean/backwashed
3. Assume added chemicals can damage the system
4. Pre-dilute the dose
   Example:
   If 1 L of acid is required
   → add 250 ml to a bucket
   → repeat 4 times
5. Add diluted chemical away from the skimmer
   (ideally in front of a return jet)
6. Repeat until complete

Commercial pools (balance tank systems)

If the pool has skimmers → use the domestic method

1. Make sure the pool is not in use
   (statutes often require this)
2. Ensure filtration is operating correctly
3. Assume added chemicals can damage the system
4. Dose in multiple locations
   (opposite scum gutter outlet)

5. Do NOT dump large amounts in one spot
6. Repeat until complete

Mixing and waiting

After dosing:

Wait at least one full turnover before reactivating the controller.

Ideally:

Wait until the next day.

Why this matters

Most chemicals have a specific gravity > 1.0, → they are heavier than water

What happens

They initially sink → uneven distribution

Result

The probe may see:

- high-concentration water
- low-concentration water

If the controller is on, it reacts incorrectly.

Before reactivation

Use monitoring only (if available).
Not dosing.

Step process

Record full water test + ORP/pH

Confirm chlorine and pH are correct

Set ORP target

Leave site

Return ≥ 24 hours later

Retest water + ORP/pH

Compare readings

If stable

Activate dosing

Observe

Watch the dosing behaviour:

Does it respond correctly?

Leave and return the next day

Retest

Adjust slightly if needed

- ±5 mV only

Repeat until stable

Tuning rules

Yes, this can be done faster.

Yes, visits can be compressed.

But:

Rules for control

1. Act intentionally
2. Adjust slowly
3. Change one variable at a time
4. Trust observed behaviour over assumptions
5. Use operator experience (where appropriate)
6. Never tune a system you do not yet understand

Simplified control rules

- Adjust slowly
- Change one variable at a time
- Wait for system response

System timing (critical)

Water does not change instantly.

What actually happens

- Initial effect → after one turnover
- Full mixing → ~6 turnovers (~95%)

Implication

A dose added today:

→ may not fully show until tomorrow

Controller behaviour must match this

If you adjust too quickly:

→ You are reacting to incomplete information

Rule of thumb

- Initial effect: ~4 hours
- Full effect: the next day or later

Last point

If this is the water's timeline:

Your adjustments must follow the same timeline.

CHAPTER 23

Setting up The Site

Explains how installation choices affect probe accuracy and controller behaviour.

ORP probes are not particularly robust. Probe behaviour does not start at the probe. It starts the moment water enters the site. There are things you put in the water, and things you let get into the water, that can affect ORP control before you even turn it on.

Chemicals Where Caution is Required

ORP probes are susceptible to poisoning, and one of the most common causes is metals. Certain products introduce metals directly into the water body and should be used with caution.

Typical examples include:

- Copper-based algaecides
- Water clarifier where the key ingredient is polyaluminium chloride
- Aluminium sulphate
- Mineral sanitiser canister (commonly copper/silver)
- Ioniser systems

Some chemicals may contain metals as an impurity. Calcium hypochlorite can contain iron impurities, which can sometimes be seen as a grey colour.

The issue with metals is not immediate. At low levels, they may have no impact on swimmer safety. However, over time, they accumulate. What begins as a trace amount gradually builds until it becomes sufficient to interfere with probe operation.

Over time, metals plate onto the platinum surface. Once that happens, readings are permanently shifted, slow, or just wrong. No cleaning fixes it. You are buying a new probe.

For this reason, it is worth considering a metal-reducing filter on the incoming mains water. Even water that is perfectly safe to drink can contain trace metals and dissolved impurities. Many times, a simple under-sink style drinking water unit is sufficient for this purpose.

Why This Matters

Water evaporates. When it does, it leaves dissolved materials behind. This is the same principle used to purify seawater through distillation. Over time, evaporation concentrates whatever is present in the water.

With metals, this means concentrations slowly increase. Eventually, the level becomes high enough to interfere with probe performance. What started as a non-issue becomes a problem simply through time and concentration.

Other Probe Fouling

Organics

Fouling is not limited to metals. A wide range of organic substances can interfere with probe performance.

General organics

Substances such as

- Sunscreen
- Body oils
- Cosmetics
- Lotions
- Hair conditioner residue
- Surfactants in clarifier
- Anti-Foaming Agent

All have a tendency to bind to surfaces. The platinum sensing surface of an ORP probe is susceptible.

These substances form a thin, often invisible film. This film interferes with electron transfer, which is exactly what ORP is trying to measure.

The result is:

- Slower probe response
- Drifting readings
- Values that do not recover properly.

Proteins

In a water body environment, proteins come from:

- Sweat
- Urine
- Saliva
- Mucus

All pools contain these to some extent.

Proteins behave differently from oils and surfactants. They tend to:

- Adhere moderately (on probes)
- Cross-link and "gel" on surfaces
- Can coat the glass membrane of pH probes

The fouling they cause is usually moderate and, importantly, often reversible. However, they still affect response time and stability.

Polymers

Polymers are one of the most significant causes of probe fouling.

They are commonly found in:

- Clarifiers
- Floc residues
- Polyquat

Polyquats (polyquaternary ammonium compounds) are long-chain molecules designed to bind particles together. A typical example is:

- poly[oxyethylene (dimethyliminio) ethylene (dimethyliminio) ethylene dichloride]

If a product contains this type of ingredient, it is polymer-based.

These products are often sold as "non-copper" algaecides, which can give the impression they are probe-safe. This is not necessarily the case. Always check the Safety Data Sheet and the manufacturer's label.

The problem with polymers is how strongly they bind. They form clear, invisible films on probe surfaces that:

- Damp the signal
- Slow response times
- Make readings appear "laggy"

Of all fouling types, polymer fouling is the one most likely to lead to probe replacement.

Clarifiers and flocculants

Many clarifiers used in the pool industry are polymer-based. Common examples include:

- Polyaluminium chloride
- Polyamines
- Polyacrylamides

Not all clarifiers contain polymers. Alternatives exist, such as.

- Natural non-polymer (chitosan)
- Enzyme-based products
- Inorganic minerals (alum)

If a clarifier must be used in an ORP-controlled system, it is essential to confirm its composition.

Similarly, many flocculants contain:

- Polyacrylamide
- PolyDADMAC
- Other polyelectrolytes

As a general rule of thumb, polymer-based flocculants should be avoided in ORP/pH-controlled water bodies unless their effects are fully understood.

Humic substances

Humic substances include:

- Leaves
- Dirt
- Plant matter
- Mulch

These introduce tannins and other organic compounds into the water.

They can:

- Foul probes
- Increase chlorine demand
- Influence pH

The fouling from humic substances is usually light to moderate and often removable with cleaning. However, if allowed to accumulate, they contribute to overall instability and demand.

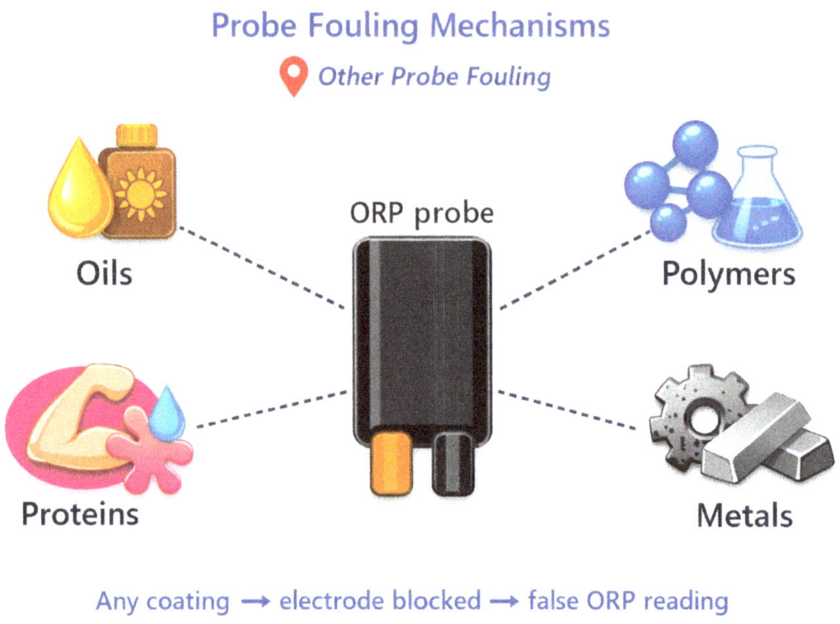

Other Interferences

Even in the absence of fouling, probe behaviour can still be unstable.

The most common cause is aeration.

Air can be introduced into the system through:

- Low pool level
- Incorrect priming
- Faulty air or vacuum release
- Leaking pump seals
- Leaking O-rings
- Cavitation
- Suction side leaks
- Variable speed changes

- Air entrapment in filters
- Pipe geometry capturing air pockets

When air reaches the probe, the readings become unstable.

System context

In industrial water treatment, systems are typically either closed or open.

Closed systems, such as chilled water loops or boiler systems, are stable. They have minimal air contact and minor variations.

Open systems, such as cooling towers or wastewater plants, are constantly interacting with the air. They are highly dynamic, with continuous chemical and biological changes.

Recreational water bodies sit between these two extremes. They are semi-open systems. They experience aeration, chemical input, biological loading, and temperature variation simultaneously. This combination makes them inherently unstable from a measurement perspective.

Probe placement

Once these factors are understood, probe placement becomes much clearer.

The probe needs to be located where the water is as stable as possible. In practical terms, this means:

- Free of aeration
- Upstream of chemical addition
- Non-turbulent
- Filtered
- pH stable

In most systems, this points to a location after the filter.

A straight run of pipe is desirable, ideally with an air release point upstream and a reasonable length of straight pipe before and after the probe.

Most times, installing the probe in the main line is preferable to a bypass. A bypass can introduce lag, meaning the probe sees changes later than they occur in the system. That said, manufacturer guidance should always be followed.

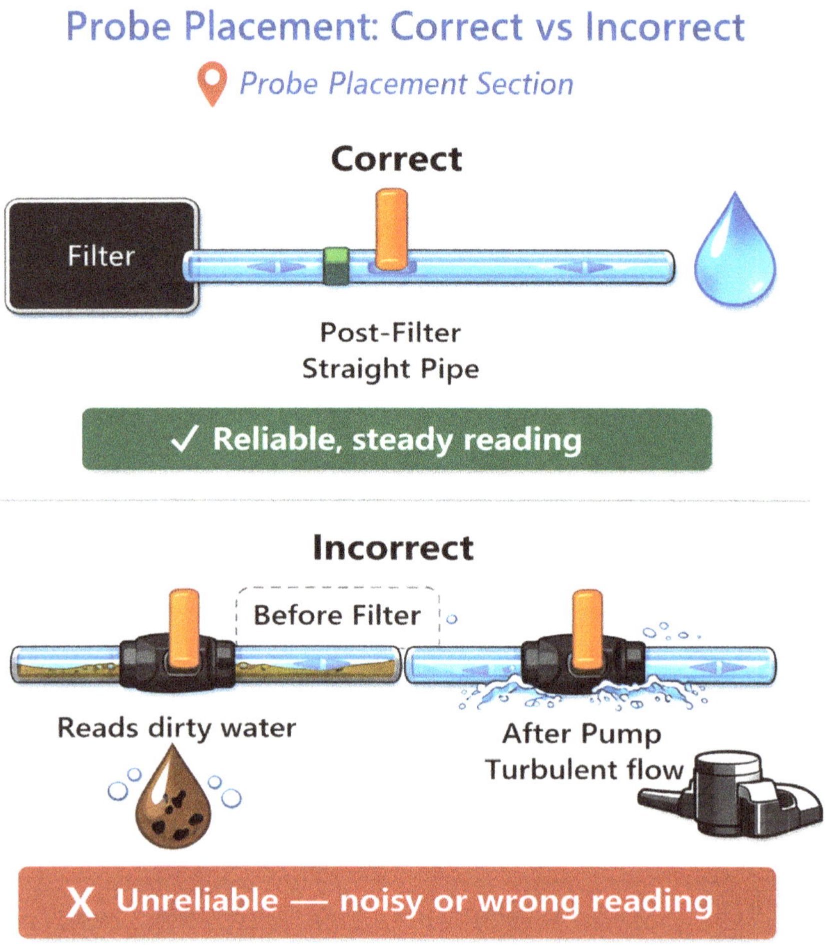

The underlying pattern should be clear. Probe behaviour is not just about the probe itself. It is a direct reflection of the environment in which it is placed, and everything from source water to chemical selection to hydraulics influences that environment.

CHAPTER 24

Chemical Storage Introduction

Introduces the basic requirements for the safe storage of pool chemicals.

The first step in any chemical dosing system, regardless of type, is the storage of the chemicals.

Typically, this may be as small as a 15–25 litre drum or carboy. Larger systems are also common, and I have designed and installed bulk storage up to 5000 litres.

Regardless of whether the system is 15 litres or 5000 litres, the same principles apply. The most important of these is the bund.

Bund

A chemical bund has specific requirements; they are:

- 110% capacity
- Fire rating
- Tan Theta

Bund capacity (110%)

The bund should be capable of holding 110% of the volume of the stored chemical.

For example:

Drum Size in Litres	Bund Size in Litres
15	16.5
20	22
25	27.5
1000	1100
2000	2200
3000	3300

This additional capacity allows for thermal expansion and is a standard approach in dangerous goods handling.

Fire rating

Fire rating requirements vary by location, so local standards should always be checked.

As a general guide, bunds often require a 2-hour rating. You may hear the term "2/2/2," which refers to:

- 2 hours holding capacity
- 2 hours external fire rating
- 2 hours flame resistance

In Western Australia, bunds for calcium or sodium hypochlorite may be plastic. This is location-specific and should always be confirmed against local requirements.

Tan Theta

One of the older principles I was taught, and still consider a sensible guide, is Tan Theta.

If a tank leaks, the liquid should still be contained within the bund.

In practical terms, the distance from the tank edge to the bund wall should be sufficient relative to the height difference, so that overflow does not escape. A common rule of thumb is that this distance should be at least 50% of the height difference.

While modern standards may approach this differently, the principle remains sound: a failure inside the bund should stay inside the bund.

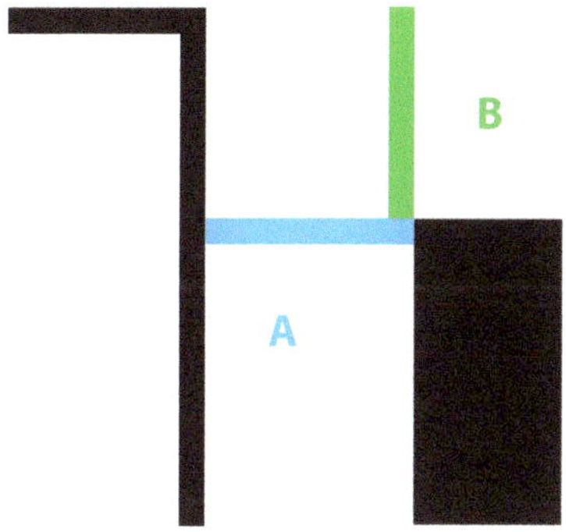

Bund sizing and 110% rule

Spill containment geometry — schematic

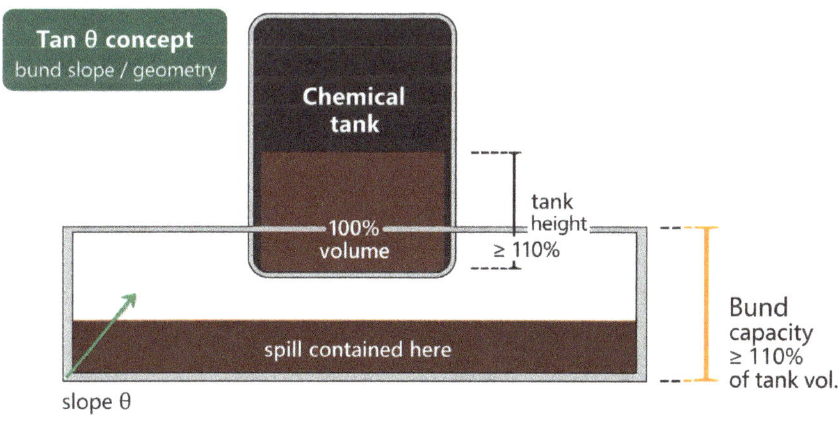

Bund **must hold** ≥ 110% of largest tank volume — full spill containment

The distance A needs to be 50% or more of the height difference B. This will prevent excess liquid from escaping the bund if a leak occurs on a full tank.

Always check local requirements, as bunds may require additional features such as low-level capture points or drainage. These are beyond the scope of this book and may require specialist advice.

Separation

Chemicals should be separated based on their classification.

If unsure, consult the Safety Data Sheet, the manufacturer, and relevant local regulations.

Separation simply means maintaining distance between incompatible chemicals.

The most straightforward way to achieve this is a physical distance within the plant room. For example, locating the acid system at one end and the sanitiser at the other, provided minimum separation distances are met.

Separation can also be achieved using structural barriers such as walls. This approach is more complex and requires input from a dangerous goods specialist.

For small storage setups, the bund itself can provide separation. If the bund is higher than the chemical container and capable of containing a spill, then effective separation can exist even with bunds placed side by side, as the contained product cannot escape.

Always ensure chemicals cannot be mixed. Clearly label chemicals and bunds to avoid this possibility.

Segregation

Segregation is closely related to separation, but focuses more on preventing accidental mixing or contamination.

In simple terms, it is about keeping chemicals apart so they cannot harm each other or the operator.

One of the simplest ways to assist with segregation is visual control. For example:

- using different coloured drums for different chemicals
- standardising bund colours across sites

For instance, all chlorine drums might be blue and all acid drums black. Even something as simple as swapping a suction line between an acid and a chlorine drum can generate hazardous fumes.

Other chemicals

Non-pool chemicals should never be stored in the same area as pool chemicals.

This includes:

- Paint
- Weedkillers
- Insecticides
- Oils
- Any unrelated chemicals

The reactions from accidental contamination are unpredictable and may be dangerous.

As a guide, maintain at least 3 metres separation, and preferably 5 metres or more.

Dry and Wet Chemicals

Ideally, dry and liquid chemicals should be stored separately.

This is not always practical, but where it cannot be avoided:

- Store unopened dry chemicals above sealed liquids
- Always confirm compatibility using the SDS

This applies to all products, including dry chemicals.

Ventilation

The term "adequate ventilation" appears in many standards, but it is not always well defined.

For indoor plant rooms, a useful rule of thumb is:

10 litres per second per square metre

This figure aligns with ventilation rates used for indoor pool environments and should manage chloramines and other fumes.

Ventilation may be natural or mechanical. Examples include:

- Open windows
- Whirlybird ventilators
- Mechanical extraction

Outdoor plant rooms rarely have ventilation issues. Indoor rooms do.

Look for indicators such as:

- Corrosion
- Sweating
- Calcification
- Salt build-up

These are signs that ventilation may be inadequate.

Example
On one site we attended, the door frame of the plant room was badly corroded. The door itself had ventilation openings, but the room was long and narrow. Heat from the equipment, combined with airflow at one end, created a stagnant, fume-heavy environment.

The result was the complete failure of the door frame within 18 months of commissioning.

Placarding

Commercial sites often require placarding.

A placard provides critical information for emergency services, including:

- Dangerous goods classification
- Chemical name
- Emergency contact details
- Relevant hazard symbols

Requirements vary between regions, so local regulations must be followed.

Placards are typically at the facility entrance. It is also good practice, and sometimes required, to place a second placard on the plant room door.

Safety Data Sheets

SDS documentation should be readily available.

Common methods include:

- A plastic sleeve fixed near the bund with printed SDS
- A labelled ring binder containing all SDS information

SDS documents should be updated regularly.

As a general rule:

- Replace every 2 years
- Or whenever the manufacturer issues an update

On larger sites, a useful practice is to request updated SDS documentation with every chemical delivery.

Summary

This section is an introduction to the storage of dangerous goods. It summarises the key principles and considerations.

For detailed design, compliance, or complex installations, specialist advice should always be sought through chemical suppliers, dangerous goods consultants, or relevant standards documentation.

CHAPTER 25

Bulk Tank Set Up

Broadly describes the structure and safety considerations of large-volume chemical storage systems.

While not directly applicable to ORP/pH control, it is common to see this type of controller on commercial properties, and many of those properties will use bulk chemical storage tanks.

This chapter is not intended as anything more than a guide to what a typical bulk storage setup looks like and what the core features are. If the design or modification of a facility is being considered, a specialist consultant should always be engaged. These installations are also state specific in Australia, and local requirements must be followed.

Bund

For bulk storage tanks, the bund is usually a walled containment area rather than a simple tray. It is typically constructed from brick or concrete and lined internally with a liquid-retaining coating. A low point is usually incorporated to assist with the clean-up and recovery of any spilled product.

A bund should be protected from the weather. It is not much use calculating 110% capacity if the bund fills with rainwater every time it rains.

Tank

Bulk tanks are most commonly used for sodium hypochlorite.

The tank material must be chemically resistant. Polyethylene is the most common choice; fibreglass can also be used, but is more expensive.

Polyethylene tanks do not last indefinitely. Sodium hypochlorite slowly degrades the material, and most times, the tank will require replacement after around seven to ten years.

Tanks are typically fitted with an access lid. This should remain closed at all times except for inspections.

Sight Glass

A sight glass is used to show the level of chemical within the tank.

The simplest form is a clear tube mounted on the side of the tank. More complex systems may use ultrasonic level measurement.

At a minimum, the sight glass should be marked at:

- Quarter
- Half
- Three-quarter
- Full

It is also good practice to include a clearly defined "call for fill" level.

Overflow

The overflow line is typically a larger pipe, often around 80 mm in diameter, above the fill point.

Its purpose is to safely carry excess liquid away if the tank is overfilled.

The discharge from the overflow is usually directed into a bucket. The bucket is filled with water so that the outlet of the overflow sits below the water level. This serves two purposes:

- Captures overflow liquid
- Helps contain fumes from the stored chemicals

The water in this bucket can be transferred to the balance tank during routine service and then refilled with tap water.

Fill Line

The fill line is typically below the overflow connection.

It should be installed so that it drains completely back into the tank after filling.

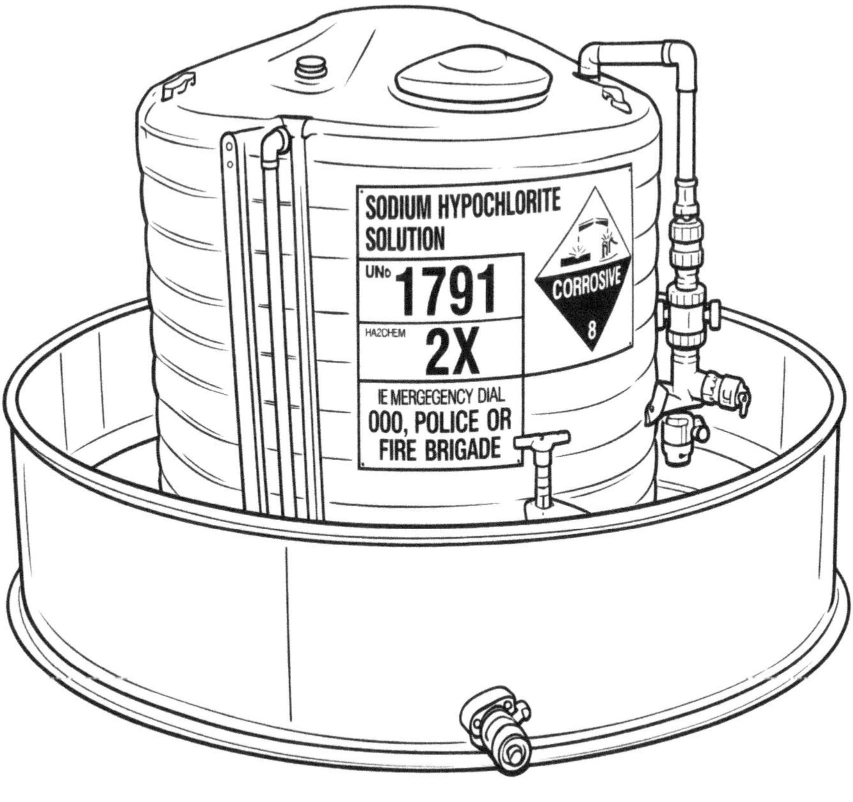

The termination arrangement must be coordinated with the chemical supplier. A camlock fitting is common, although this varies by region and supplier.

When not in use, the fill line should be capped.

Take Off assembly

Most systems include a dedicated take-off arrangement to supply the dosing equipment.

This is usually a rigid PVC pipe setup. The typical sequence of components is:

1. Isolation valve—allows shutdown of the supply
2. Strainer—captures debris or swarf from the tank
3. Isolation valve—allows the strainer to be serviced
4. Pipe run—feeds the dosing system
5. Gas vent—vertical pipe extending above the tank level
6. Take-off valve—final control before dosing equipment

This arrangement allows the system to be isolated, cleaned, and serviced without shutting down the entire installation.

Safety Shower

A safety shower (or deluge shower) should be within a few metres of the filling point.

Specific requirements such as location, flow rate, and coverage will depend on local standards and must be verified.

Warnings

This type of installation falls under dangerous goods storage and should not be treated lightly. This chapter is not a substitute for professional advice.

There are several critical points worth reinforcing:

PVC solvent cement is not compatible with many pool chemicals. It will fail quickly, and often without warning. Specialist chemical-resistant adhesives must be used.

All UPVC pipes used for chemical service should be at least Class 18. CPVC Schedule 44 is typically acceptable, but always confirm with the supplier.

All valves and fittings should use Viton or equivalent seals and materials suitable for chemical exposure.

Do not repair leaking tanks, valves, or fittings unless properly trained. Bulk chemical storage systems require specialist knowledge, and incorrect repairs can cause serious hazards.

CHAPTER 26

Tubing Installation

Explains how chemical transfer lines should be installed to avoid operational problems.

For readers following from the beginning, I apologise in advance. This material was covered in my other book, Carbon Dioxide for pH Control. Parts of that chapter reappear here, with the appendix on painting conduit copied verbatim from that text. I recommend reading that book if you have not already done so.

Suction lines should remain short wherever practical, and discharge lines can be longer to suit. While not entirely correct for chemical dosing, it is still a good practice.

Fortunately, for sodium hypochlorite and acid dosing, the dosing pumps will come as a complete kit.

The kits typically include:

1. Foot valve
2. Suction tube
3. Dosing pump
4. Discharge tube
5. Injector

To keep the foot valve submerged in the drum, it may require weighting. One of the best weights for this is an electrical ceramic insulator.

Take the tube and foot valve to an electrical parts supplier and ask them for a ceramic insulator that fits over the tube. If it fits through the lid of the drum, this is an excellent choice.

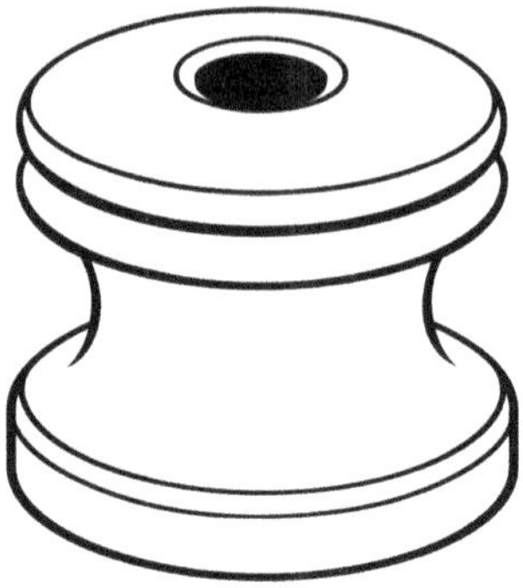

Other methods can include using a short section of PVC to create a holder. Letting the foot valve float within the liquid and prime as needed, hoping the primed tube suffices to hold the foot valve where needed.

Connecting to pump

For diaphragm pumps, the recommended procedure is to guide the tube upwards, form a relaxed loop, and then connect it to the pump's suction. The loop helps reduce strain on the tube, as the pulsation of some dosing pumps can be severe.

Identifying chemicals

In a domestic setting, it is not usual to encase the dosing tube. On commercial sites, however, it is best practice to do so.

Electrical conduit is small enough for a dosing tube and available in both rigid and flexible, making it useful for the purpose. You should paint the conduit the colour of the substance the tube transfers. See Chapter 52, Appendix Painting Plastics for the painting method.

In Australia, the colours of this encasement are determined by AS 1345–1995–Identification of the Contents of Pipes, Conduits and Ducts. Under this standard, sodium hypochlorite or calcium hypochlorite is violet, as is acid. The official colour code is Lilac P23.

As both chlorine and acid are in similarly coloured conduits, it is best practice to label the conduits to show:

- The contents
- Direction of flow

A simple vinyl permanent sticker will work well.

Trim the arrow from the end that is not in the direction of flow.

CHAPTER 27

Dosing Pumps/ Solenoid

Compares different dosing devices and their effects on chemical control.

As previously mentioned, there are several types of dosing devices commonly used in aquatic facilities:

- Venturi
- Solenoid valves (valves that open and close electrically)
- Peristaltic pumps (roller pumps, squeezing liquid through a tube)
- Diaphragm pumps (pumps using a pulsing diaphragm to move liquid)

Venturi Systems

The least controllable option is the venturi. In the early 1990s, a common approach involved a simple plastic tapping band that created a venturi vacuum and used a rotary knob to adjust flow. These systems were inexpensive, inconsistent, and short-lived—often lasting only a single season before being discarded. Their chief appeal was cost, not performance.

Solenoid-Based Dosing

Solenoids are still used, particularly in domestic pools where flow rates are modest and the level of control required is not as stringent.

Most solenoids have:

- A minimum pass-through rate, and
- A minimum open time

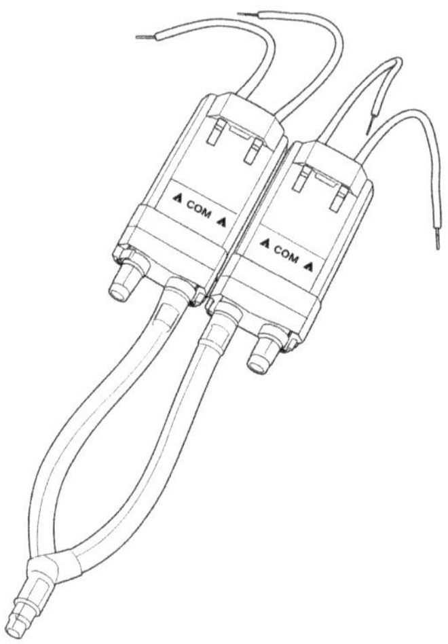

This makes them suitable for pools around 50,000 litres and above, where minimum flow and turnover conditions allow proper function.

Peristaltic Pumps

A step above solenoids, peristaltic pumps provide more reliable and controllable dosing.

They can be controlled digitally using on/off logic, which is typically referred to as digital or on/off control.

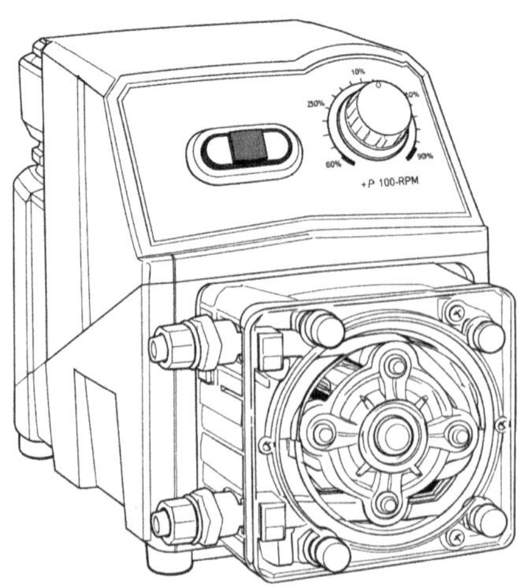

A dosing pump wired directly to mains power and switched on/off by a controller is an example of this method.

Control Methods Overview

This book focuses primarily on ORP/pH control, but an overview of common dosing-control signals is helpful. Control theory is a broader technical subject, and specialised texts should be consulted for more depth.

Analogue vs Digital Control

Analogue control refers to any non-digital signal type.

Common analogue methods include:

- 4–20 mA signals
- Pulse signals
- Other proprietary or variable-frequency outputs

4–20 mA Control

This is one of the most common dosing-control methods:

The controller sends a variable current to the dosing pump.

- 4 mA typically represents the minimum output (or zero/stop).
- 20 mA represents the maximum output.
- Some pumps will stop above 20 mA as a safety behaviour.

The pump adjusts its feed rate proportionally to the signal strength.

Pulse Control

Pulse signals vary by manufacturer, but operate in one of two ways:

- Direct pulsing—each pulse triggers one physical pump stroke.
- Pulse-counting logic—the pump counts incoming pulses and uses internal algorithms to determine dose timing and rate.

Both methods allow the dosing controller to vary output based on the frequency of transmitted pulses.

Diaphragm Pumps

Diaphragm pumps are more typically found on larger commercial or industrial sites. This type of pump is more likely under analogue control.

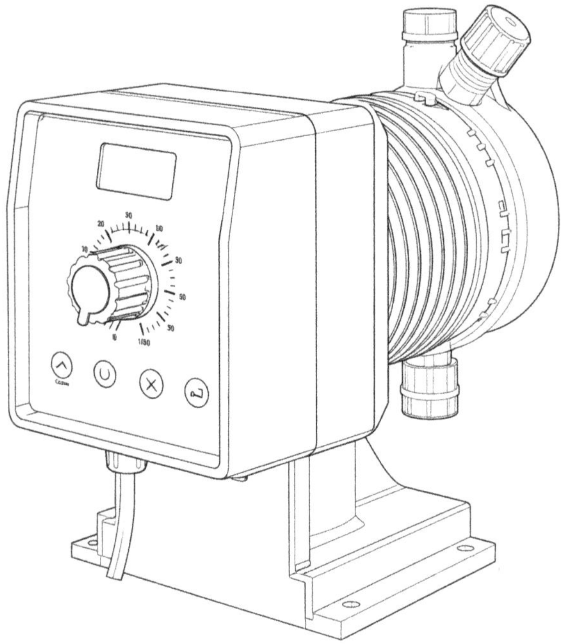

These pumps are favoured as they can adjust their stroke length and stroke speed. This gives them a much wider range of dosing amounts.

They are usually the more expensive option, usually at least four or five times the cost of a peristaltic pump.

These pumps have additional features and control devices. If this is installed, it is best to consult with the manufacturer's agent before adjusting it.

Diaphragm pumps can produce a very solid slug of liquid. For this reason, it is best to loop both the inlet and outlet tubes. It provides stress relief for the tube by allowing the vibration to smooth out before transfer to the injection point.

CHAPTER 28

Injector

Explains how chemicals are introduced into the system and why this affects dosing accuracy.

For most domestic installations, the injector is a small check valve and some form of tube attachment.

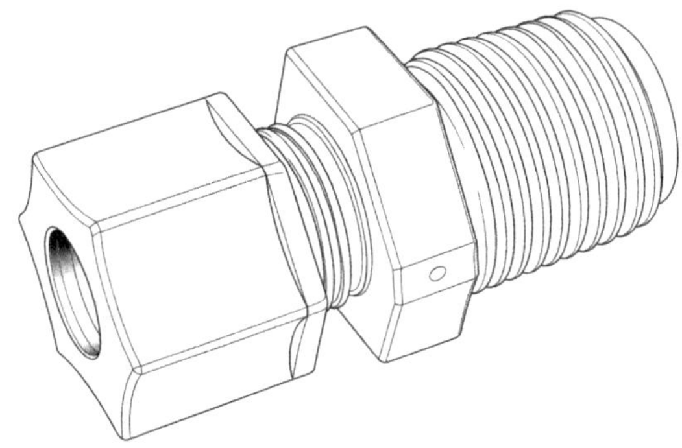

For more sophisticated sites, an injection lance or a withdrawable injector is commonly used.

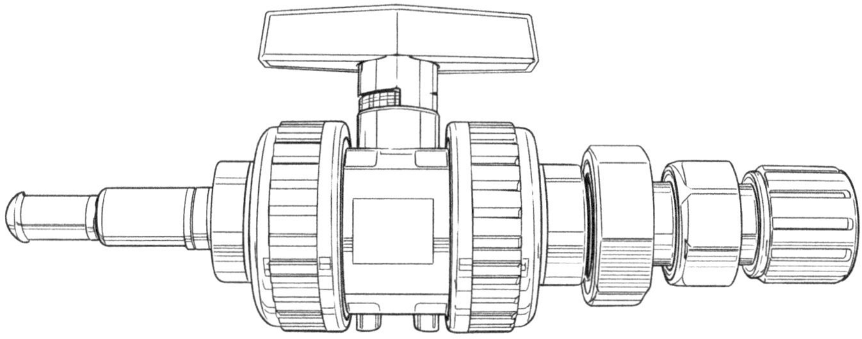

Some manufacturers will combine acid and chlorine into one injector. Mixing acid and chlorine can form chlorine gas, and this type of injector should be approached with caution.

If a mixed injector is present, then each chemical line must be fully flushed with water before attempting to clean. Never:

- Flush both lines at the same time
- Mix the chlorine and acid lines
- Flush acid with chlorine
- Flush chlorine with acid
- Flush both lines from the same container of water

Even a small amount of acid mixing with chlorine can cause harm.

Before working on the injectors:

- Review the safety data sheet for the product
- Locate deluge shower, eyewash, or fresh water supply
- Don appropriate personal protective equipment
- Flush the line with water
- Pull the suction end out of the chemical or isolate the chemical feed line.
- Relieve any pressure in the dosing line as per the equipment manufacturer's instructions

Once the injector is isolated, flushed, and safe to work on, remove it from the pipeline according to the manufacturer's instructions.

Depending on the chemical being dosed, the buildup is more typically calcium-based carbonate products. Some buildup will be sodium chloride.

Injector Cleaning

Injector cleaning should proceed on a least-to-most aggressive basis. Always try the least aggressive cleaning first and only escalate if needed. Always cleaning with the most aggressive first will cause a shortened life of components.

Step 1 — Rinse with clean water

Mildest: System flushing with fresh water

Purpose: Clear loose debris, crystallization, or salt flakes without chemicals.

Method:

- Disconnect one chemical line at a time (chlorine or acid).
- Place the suction line into a bucket of clean water.
- Use the controller's manual dose function to pump fresh water through the line.
- Repeat on the other line separately.

Why:

This clears most blockages without chemical exposure and avoids

mixing chlorine and acid inside the injector.

Step 2 — Soak in warm water

Remove injector fittings and soak in warm water

If flushing is not enough:

Method:

- Unscrew the injector(s) from the manifold.
- Soak the fitting(s) in warm water for 10–20 minutes.
- Use a soft plastic cable tie to poke out the scale (never metal).

Why:

Warm water softens hardened chemical crusts (especially sodium hypochlorite crystals) safely.

Step 3 – Dilute acid soak

Slightly more aggressive: Mild acid soak (acid injector only)

ONLY for the acid-injection fitting, and never anywhere near chlorine components.

Method:

- Soak the acid injector in a very mild acid solution (10:1 water: acid).
- Rinse thoroughly afterwards.

Why:

Calcium deposits and scale dissolve in a weak acid.

More aggressive: Replace non-return valves or injectors

If buildup or backflow has deformed valve seats or duckbills:

Method:

- Replace the injector or the non-return valve.
- This is usually inexpensive and restores proper flow.

Why:

Chemical wear eventually degrades internal elastomers. Replacement is safer than over-cleaning.

Step 4 – Mechanical cleaning

Most aggressive: Manifold removal and descaling (rare)

Only performed when:

- Injection manifold ports are blocked, or.
- Internal corrosion, scale, or plastic deformation is clear.

Method:

- Disassemble the manifold.
- Flush with water.
- Acid-soak only if scale is present, and only after all chlorine components are removed from the workspace.
- Replace if significantly worn.

Why:

Restores full flow when all other measures fail.

Prevents uneven dosing or chemical backflow.

Step 5 – Replace injector

Critical safety note

1. Always review the safety data sheet for the chemical being dosed.
2. Always don all protective personal equipment as detailed in the SDS.
3. Never mix chlorine and acid components during cleaning.
4. Never attempt to clear injectors using compressed air (risk of blowback or chemical aerosolization).

5. Use only plastic tools to avoid damaging the injector bores.
6. Always flush lines individually, even when a system uses a combined manifold.

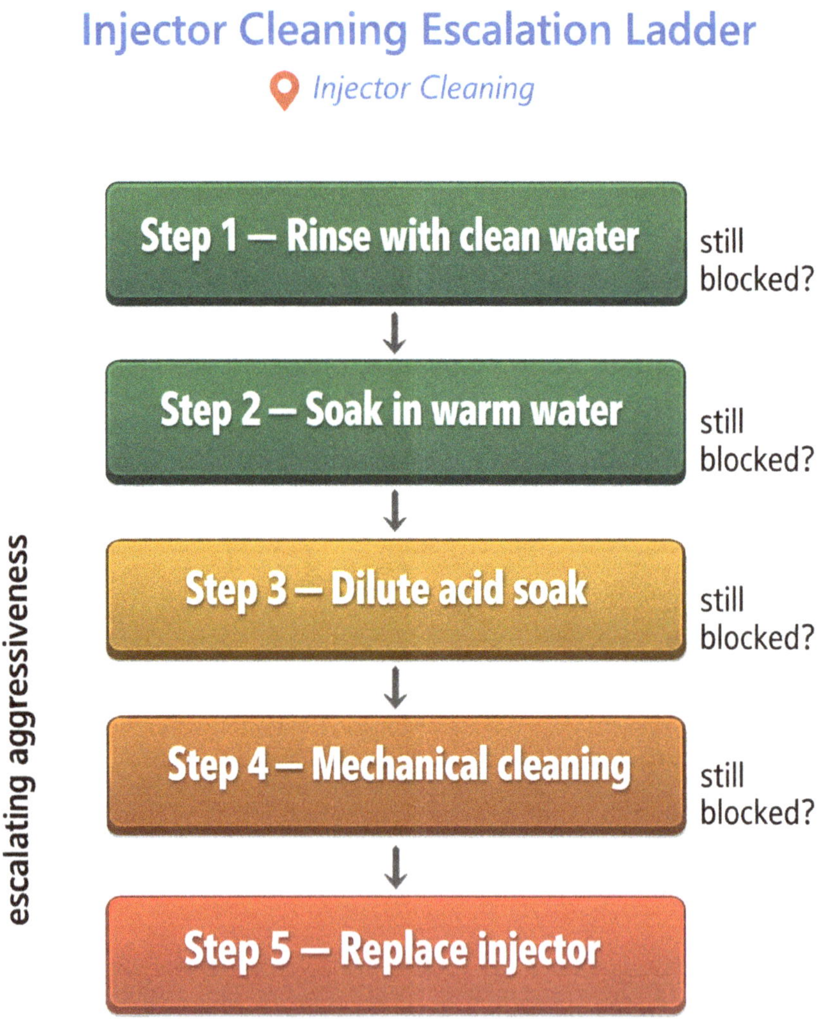

Least — most aggressive cleaning decision path

CHAPTER 29

What the Controller is Actually Doing

Explains the simple logic of the controller and what it does not know about the system.

This chapter is about the box on the wall. Not the probe, not the dosing pump — the controller itself. Most operators can describe what an ORP controller measures. Very few can describe what it decides, how it decides it, and what it cannot see.

That gap matters. A controller that works incorrectly looks almost identical to one that works correctly. Knowing how the logic runs is the only way to tell them apart.

The Control Loop in Plain Terms

An ORP/pH controller runs a simple, continuous or semi-continuous comparison:

Is the probe reading above or below the target value?

If the reading is below the target, the controller signals the dosing device. If the reading is at or above the target, it stops.

That is the entire decision. The controller does not know your pool volume.

It does not know your bather count.

It does not know what chemical you loaded into the drum, when you last calibrated the probe, or whether your filter is clean.

It knows one number. The millivolt reading the probe is currently producing, and it compares that to the number you programmed in as a target.

Everything else in this book — the chemistry, the hydraulics, the CYA effect, the temperature drift — matters precisely because it affects that single number reaching the controller. The controller itself has no visibility into any of those factors. It receives a voltage. It acts on the voltage. A fouled probe, a pH that has drifted, and a CYA level. From the controller's perspective, all look identical. They are all just a lower number.

Deadband and Pool Chemistry Are Not Independent

The deadband (range where the controller does nothing) setting looks like a simple control parameter. It is not. Its real-world effect depends entirely on where you are on the ORP curve at the time of dosing.

At low free chlorine levels, say 0.5 to 1.0 mg/L in an unstabilised pool, the ORP response to a chlorine addition is steep. A small chemical addition produces a large millivolt movement. A 20–30 mV deadband in this region could represent a modest and predictable chlorine addition.

At higher free chlorine levels, the relationship changes, at 3–4 mg/L free chlorine with pH around 7.5, some controller algorithms have been observed to produce chlorine jumps from 4 mg/L to above 10 mg/L in response to a 20–30 mV correction cycle. The deadband has not changed. The pool chemistry has.

This happens because of two compounding effects.

First, the ORP curve flattens at higher chlorine concentrations. The same millivolt movement that represented a modest chlorine addition at 1 mg/L now represents a much larger one at 4 mg/L, because you are adding into an already-oxidised system and the probe is responding to the combined redox environment, not just the incremental addition.

Second, the dosing device continues to operate during the chemical transit time between the injection point and the probe. The controller signals to dose until the probe recovers to the upper deadband limit. But the chemical already added has not reached the probe yet. The controller does not know this. It keeps dosing into the lag. By the time the dosed chemical arrives at the probe, and the reading climbs above the deadband, the pool has received substantially more chemical than the controller intended to deliver. This is another reason these controllers are less suited to smaller water bodies.

The combined result is overshoot, sometimes severe overshoot, in pools where the chemistry is already at a functional level. The controller was not malfunctioning. It was doing exactly what it did. The design assumption was that a 20–30 mV band represents a consistent chemical dose. It may not.

The practical implication is this: a deadband that is appropriate at startup, when the pool is being brought to target from a low level, may be inappropriate as a steady-state operating parameter once the pool is at or near target. Some controllers may allow separate configuration for the initial ramp-up phase and the maintenance phase. Where this is available, it should be used. Where it is not, the operator needs to know that the controller's behaviour in maintenance mode is not the same as its behaviour during correction, and that the manual test is the only tool that confirms what the chemistry is actually doing.

In practice, I would rather start up the pool using manual dosing wherever possible and then tune the ORP/pH control to the final setpoints I have achieved manually. It may not look as sophisticated as using automation, but I will accept reliability over appearance every time.

This is also why the guidance in the Chapter 22 Chemical Dosing with an ORP Controller Installed — adjust the setpoint by only 5 mV at a time, wait at least one full turnover, test fully before adjusting again — is not conservative caution. It is the correct response time for a system whose output is non-linear and whose effects are delayed and sometimes severely delayed.

What the Controller Does Not Know

This is worth stating plainly because it is the most common source of misplaced confidence in ORP/pH systems.

The controller does not know whether its probe is fouled. A probe coated with a film of polymer or body oil will produce a sluggish, suppressed signal. The controller receives that signal and acts on it exactly as it would act on a clean, accurate signal. It will dose more because the reading is low. The reading is low not because the water needs more chemicals, but because the probe cannot see the water correctly. The water may already be well-dosed or overdosed.

> **Rule:** *The controller trusts the probe.*
> *Even when the probe is wrong.*

The controller does not know whether the probe has drifted out of calibration. Probes drift. The reference electrode inside the probe ages. A probe that read accurately six months ago may now read 40 mV low. The controller has no way of knowing this. It will simply treat every day as if the probe is correct.

The controller does not know about CYA. As covered in earlier chapters, cyanuric acid suppresses the ORP signal substantially. A pool at 50 mg/L CYA will produce a meaningfully lower ORP reading than a pool at the same free chlorine level with no CYA. The controller sees a lower number and doses accordingly. It has no awareness that the suppression is chemical rather than a genuine deficit of sanitiser.

The controller does not know about pH. This deserves emphasis because pH has a direct, large effect on the ORP reading, roughly 50–60 mV per pH unit, as described in Chapter 18 in the Nernst equation. If the pH control side of the system is not working correctly, the ORP reading will move with pH, and the controller will respond to pH changes as if they were changes in sanitiser level. An unstable pH means an unstable ORP means an unstable controller, not because the dosing logic is broken, but because the input it is working from is unreliable.

Rule: *The controller cannot separate chlorine from pH.*

The controller does not know the dilution rate of new chemicals, nor the amount of new water being added. When the controller triggers a dose, it does not know whether the chemical that arrives at the probe site is the same concentration you intended. A degraded drum of sodium hypochlorite (covered later in Chapter 39 Chemical Quality and Degradation) delivers less active chlorine per millilitre than a fresh drum. From the controller's perspective, it dosed. If the probe reading did not move adequately, it doses again. The system will continue dosing at an elevated rate to compensate for a degraded chemical, unless the operator catches it manually.

How the Controller Fails as a Dosing Device

The controller fails in two directions: it overdoses, or it underdoses. Both are dangerous, and both can appear stable on the controller display.

Overdosing typically occurs when the probe is producing a suppressed or sluggish reading. The controller interprets the low reading as insufficient sanitiser and continues dosing past the point of adequacy. In small volumes, spas, and small commercial pools, this can produce chlorine levels well above safe limits before the error is caught. This is not a theoretical risk. It is a failure mode that has caused injuries. The spray-pad incident described earlier in this book is an example of exactly this pattern.

Overdosing can also occur when a significant volume of fresh water is added to the pool. Fresh makeup water arrives at the probe at a lower chemical concentration than the bulk pool water. The controller reads this as a chemistry deficit and doses even though the bulk pool may have been at target. The practical implication for the control loop (sense → decide → dose cycle) is that the controller has no averaging or delay logic to handle the dilution pulse. It sees a low number and responds immediately.

Underdosing typically occurs when the probe is producing a higher reading that does not reflect the true water condition. A probe that has been contaminated by a substance with a high redox potential, certain metals, for example, can read artificially high.

The controller, seeing a high reading, doses little or nothing.

The actual sanitiser level falls.

The water is unsafe, while the controller displays a satisfactory reading.

Underdosing also occurs when the dosing device fails mechanically:

- a blocked injector,
- a failed non-return valve,
- a peristaltic pump tube that has cracked.

While the probe continues to report a falling reading. The controller keeps signalling to dose, but nothing is arriving at the water.
This will eventually drive the controller to maximum output, still delivering nothing.

The Controller Is Not the System

This is the point.

An ORP/pH controller is a responsive device. It reacts to what the probe reports. It does not model the pool. It does not predict demand. It does not account for:

- bather load,
- weather,
- top up water inflow,
- last addition of chemicals,
- time of day,
- or chemical degradation.

It compares one number to another number and switches a relay.

> **Rule:** *The controller only compares a number and reacts. It does not understand the pool.*

This architecture makes it useful in stable conditions.

It makes it unreliable in variable ones.

A stable indoor pool with controlled temperature, predictable chemistry, a well-maintained probe, and consistent bather loading is the environment in which the controller works. Most real pools deviate from this in one or more ways, some of the time.

The manual test does not exist to duplicate what the controller is doing. It exists to check whether what the controller is doing corresponds to reality. These are different tasks. The controller's display confirms that the controller's probe is producing a reading in the right range. The manual DPD test confirms whether the water is safe. These can agree, or they can disagree, and they disagree often enough that treating manual testing as redundant has caused harm.

The controller is a capable assistant under appropriate conditions; the responsibility for water safety rests with the operator. These two things do not contradict each other; they define the professional relationship between an operator and their equipment.

Using the Reference Tables to Cross-Check Controller Behaviour

The reference tables in the appendix give approximate expected ORP values at known combinations of free chlorine, pH, and CYA level. They are compiled from field observations and published datasets, not from theoretical predictions alone.

Their most useful application is not lookup; it is cross-checking.

When you do your manual test, you now have free chlorine, pH, and CYA. The tables tell you approximately what ORP the probe may read, given those parameters. Compare that to what the controller is actually displaying.

If the controller reading is broadly consistent with what the table predicts, the system is behaving. If there is a significant gap, say, 40 mV or more in either direction, something is different and worth investigating.

The probe may need cleaning or calibration.

The chemistry may have a variable that the tables do not account for.

The dosing system may have been working on a degraded chemical.

The investigation starts there, not with the controller's setpoint.

This cross-check does not require any calculation. It requires a manual test, a table lookup, and a comparison. It converts the appendix from reference material into a practical diagnostic tool, and it gives the operator a method for auditing the controller's behaviour rather than simply trusting the display.

Look, sometimes a specific install will read completely differently for some peculiar reason that remains shrouded. I assume it is likely a minor chemical difference, or perhaps it is as simple as the amount of sunlight reacting with the water in the skimmer box; it is a mystery. Maybe the difference can be accepted and the water body successfully managed within this reading; however, at other times, the site is better served with an alternative technology.

CHAPTER 30

ORP Probe Cleaning

Sets out how to maintain probes to keep readings reliable.

Like injector cleaning, probe cleaning should progress from mildest to most aggressive, unless the specific cause of fouling is known.

Rules for probes:

1. Never directly touch the electrode with fingers—finger oils can damage the probe's sensors.
2. Never scratch, scrub, or use anything abrasive on the probe—the platinum and glass can be damaged.
3. Never store probes dry (unless explicitly "dry-storage" type)—most probes need to be stored in potassium chloride (KCl) solution, never in water or demineralised, deionised, or any other type of water. The probe's internal salts can be destabilised.
4. Rinse with clean water after removal, but wipe only with a soft cloth—dab the probe dry, do not wipe.
5. Never allow chlorine or acid to sit on a probe tip.
6. Calibrate regularly and only use fresh buffer solutions.
7. Protect the probes from UV.
8. Protect the probes from electrical interference.

Over the next few pages, suggested cleaning regimens are provided. These are generic and are not intended to replace manufacturers' instructions. If available, always refer to and follow the manufacturer's instructions.

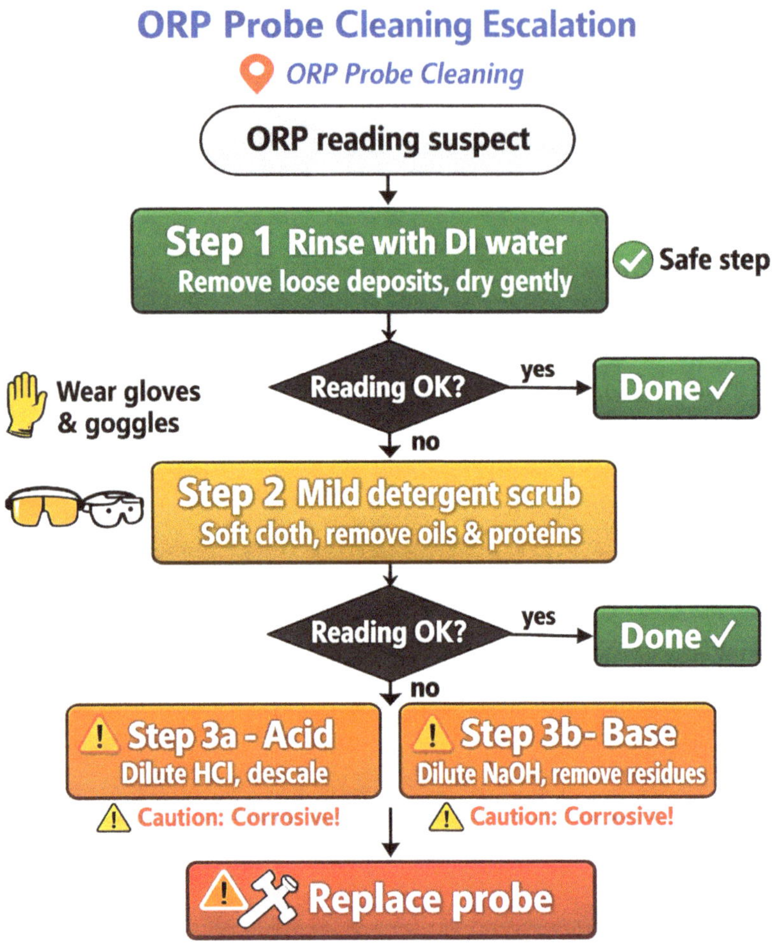

Decision tree: rinse → detergent → acid/base → replace

Freshwater Rinse (Least Aggressive)

Purpose: Remove loose debris, light contamination, and surface films.

Required Items:

- Plastic cup or similar
- Distilled water
- KCl solution
- pH 4 buffer

Method:

1. Isolate probe.
2. Remove from the pipework.
3. Rinse the probe under clean water (preferably distilled).
4. Gently wipe around the body with a soft cloth, not the sensing element itself.
5. ORP probe soak in KCl for about 2 minutes.
6. pH probe soak in pH 4 buffer for about 2 minutes.
7. If calibration is required, recalibrate—refer to the manufacturer's instructions.
8. Return to pipework.
9. Return to service.

Expected Outcome:

Readings return to the normal pattern, and dosing resumes. If not, escalate to the next step.

Why:

Basic rinsing is the first recommended step in standard probe care guidance.

Mild Detergent / Surfactant Cleaning

Purpose: Remove organic films, oils, greases, sunscreen residues, or biofilm.

Required Items:

- 2 or more plastic cups or similar
- Distilled water
- Mild, non-toxic, non-ionic dishwashing liquid
- KCl solution
- pH 4 buffer

Method:

1. Isolate probe.
2. Remove from the pipework.
3. Rinse the probe under clean water (preferably distilled).
4. Using the dishwashing liquid, add a few drops to the water and mix thoroughly. Then soak the probe for 5 to 15 minutes.
5. Rinse the probe thoroughly afterwards.
6. Gently wipe around the body with a soft cloth, not the sensing element itself.
7. ORP probe soak in KCl for about 2 minutes.
8. pH probe soak in pH 4 buffer for about 2 minutes.
9. If calibration is required, recalibrate—refer to the manufacturer's instructions.
10. Return to the pipework.
11. Return to service.

Expected Outcome:

Readings return to the normal pattern, and dosing resumes. If not, escalate to the next step.

Why:

General electrode maintenance guidelines approve surfactants for organic contamination removal.

Targeted Chemical Cleaning (Moderate Aggression)

Use only when detergent cleaning has not restored stable readings.

Required Items:

- 2 or more plastic cups or similar
- Distilled water

- 5-15% hydrochloric acid—the best idea is to purchase this from a lab or lab supply
- 5–15% sodium hydroxide–the best idea is to purchase this from a lab or lab supply
- KCl solution
- pH 4 buffer

Method:

For mineral / alkaline deposits:

1. Isolate probe.
2. Remove from the pipework.
3. Rinse the probe under clean water (preferably distilled).
4. Clean with 5–15% hydrochloric acid <30-second soaks.
5. Rinse the probe thoroughly after each soak.
6. Gently wipe around the body with a soft cloth, not the sensing element itself.
7. ORP probe soak in KCl for about 2 minutes.
8. pH probe soak in pH 4 buffer for about 2 minutes.
9. If calibration is required, recalibrate—refer to the manufacturer's instructions.
10. Return to the pipework.
11. Return to service.

For heavy organic contaminants:

1. Isolate probe.
2. Remove from the pipework.
3. Rinse the probe under clean water (preferably distilled).

4. Clean with 5–15% sodium hydroxide<30 second soaks.
5. Rinse the probe thoroughly after each soak.
6. Gently wipe around the body with a soft cloth, not the sensing element itself.
7. ORP probe soak in KCl for about 2 minutes.
8. pH probe soak in pH 4 buffer for about 2 minutes.
9. If calibration is required, recalibrate—refer to the manufacturer's instructions.
10. Return to the pipework.
11. Return to service.

Expected Outcome:

Readings return to the normal pattern, and dosing resumes. If not, escalate to the next step.

Why:

Industrial probe cleaning standards specify acid for alkaline/mineral scale and sodium hydroxide for organic contamination.

Replacement (Most Aggressive Step of All)

If the probe still:

- will not calibrate,
- drifts excessively, or
- shows a very slow response time.

It is likely exhausted and should be replaced.

Most probes have a typical lifespan of 1–3 years, depending on water conditions.

CHAPTER 31

Mothballing and Probe Storage

Explains how to safely shut down and store dosing and probe systems.

Before COVID, this chapter would likely not have existed, as the need arose so infrequently it was not worth mentioning. During COVID, we were mothballing systems everywhere and moving to manual dosing.

To mothball a system for an unknown period. Choices include:

1. Run the plant as normal and make no changes.
2. Run the plant as normal and reduce the chlorine target, perhaps increase pH as well.
3. Run the plant less with chemical control in place.
4. Run the plant less without chemical control and move to manual dosing.

There will be others as well.

The choice we made in April 2020 was choice 4. We did not know if we were closing the pools for a week, a month, a year, or forever.

In this chapter, I will summarise the steps to mothball the chemical dosing plant.

The steps to take are:

1. Run the chemical drums down to empty.
2. Flush all dosing pumps and lines with water.

3. Thoroughly clean all injectors, including stripping down and cleaning of internal components (according to the manufacturer's instructions)
4. Open all dosing lines and drain dry.
5. Open up diaphragm pump heads and drain dry (all our systems were diaphragm).
6. Isolate the water flow to the probes.
7. Remove probes.
8. Drain water from probe lines if needed.
9. Store probes in a KCl solution or according to the manufacturer's instructions.

Once the chemical dosing is mothballed, the plant can be run for a reduced time if water is being held in the water bodies.

If the water bodies are to be emptied (please check with a pool builder before emptying any water body), then the equipment and pipework can be drained down as well.

Probes stored correctly can last several years. Refer to the manufacturer's instructions for specific instructions for their equipment.

CHAPTER 32

Chlorine has a job to do. What is it

Defines the role of chlorine as a disinfectant and oxidiser in pool water.

Chlorine can sound confusing as a sanitiser and oxidiser. How can it be two at the same time?

Try thinking about it this way.

Chlorine sanitises through oxidation.

The hypochlorous acid is the oxidation component of chlorine. This is the part that is active and is a sanitiser. The hypochlorite ion also sanitises; however, is not a strong oxidant.

Combining with organics will cause the hypochlorous acid to be reduced. This degradation of chlorine can be significant.

Maybe a water body owner may request a new controller due to not being able to maintain chlorine in the water.

Always check the potential for organic contaminants driving consumption to higher or unsustainable levels.

Algae growth is a common visual impact of lower sanitiser levels. The growth can be localised to a specific spot, which is an indicator of a dead water area, or it can be more general, which is an indicator of other issues.

Other issues could be:

- Low or slow turnover
- Insufficient turnover time
- Dirty filters
- Overdue media change

Other potential issues may exist as well on-site and can only be determined by a thorough site inspection. Inspect the filtration plant first before assuming chemical faults, although, of course, chemicals, or lack thereof, may be the main cause.

The sanitiser has a higher calling. Not only is its purpose to oxidise organics and control algae, but it is also there to protect human health by sanitising the pathogens associated with recreational water illnesses.

For a more thorough examination of recreational water illnesses, please see my pending book, Recreational Water Illnesses—a Pool Bloke's Guide.

Pathogens may be introduced to the water from:

- Environment.
- Blown-in litter.
- Bathers' bodies.
- Fill water.
- Run off.
- And a host of other vectors.

Demand from pathogen-associated incidents is usually minor. However, there is one incident where demand spikes exponentially.

Any incident involving faecal contamination from any source can be a significant demand generator.

Chemical controllers function to maintain chemical levels, and while they will respond to a one-off incident. That is not their function.

Manual chemical intervention is required from time to time.

CHAPTER 33

Manual Chemical Intervention

Explains when and how manual dosing is required despite automation.

Regardless of the chemical controller, from time to time, manual chemical intervention is required.

Some basic rules for chemical interventions apply to nearly all chemical controllers. Before chemical dosing:

- Isolate the probes.
- Follow the chemical label instructions. If mixing is required, the product must be mixed.

I am assuming the filtration is running.

The isolation of the probes prevents them from being potentially damaged by concentrated chemicals and helps keep calibration.

Following chemical label instructions is important all the time; however, more important on a chemical controller pool than on others. Mixing a chemical with water before addition to the pool assists in reducing the possibility of a slug of chemical hitting the probes or lodging in the water circuit leading to the probes and causing issues later.

Wait at least one turnover and test the water fully before de-isolating the probes. The water must return to stability before the probes are enabled.

Some chemicals, such as cyanuric acid, that take time to dissolve may need the wait time to be longer. If I had to wait a week, I would not mind. Better to wait a week and protect the equipment than rush and damage something unexpectedly.

CHAPTER 34

ORP and Salt

Explains why salt systems and ORP control interact poorly many times.

This combines two areas I use less than a lot of others. As most of my work is commercial and the specific regulatory framework I work in requires every salt pool to also have an independent chemical dosing system, there is little benefit in fitting salt. It simply adds a cost.

Domestically, a few years ago, I noted a lot of salt chlorinators on the market with ORP/pH controllers integrated. Recently, I thought I needed one and could not find any available. Why?

Saltwater chlorination is one area bound to cause disagreement. Some people will say it generates chlorine gas, some will say it produces sodium hypochlorite, and others will say it does not, as there is no hydroxide.

Saltwater chlorinators convert sodium chloride (NaCl) dissolved in pool water into free chlorine during electrolysis. The full set of reactions is:

1. Electrolysis of Salt

Electrolysis is the use of electricity to cause chemical reactions.

At the anode (positive electrode):

$$2Cl^{-} \rightarrow Cl_2(g) + 2e^{-}$$

At the cathode (negative electrode):

$$2H_2O + 2e- \rightarrow 2OH- + H_{2(g)}$$

2. Chlorine Dissolves in Water

The chlorine gas generated dissolves immediately:

$$Cl_2 + H_2O \rightarrow HOCl + H^+ + Cl^-$$

3. Formation of Free Chlorine (the sanitiser)

The key sanitiser in pool water is hypochlorous acid (HOCl), which partly dissociates:

$$HOCl \rightleftharpoons H^+ + OCl^-$$

Put Together: The Common Overall Chlorinator Equation

Combining the anode, cathode, and hydrolysis steps gives the simplified "textbook" salt chlorinator equation:

$$2NaCl + 2H_2O \rightarrow Cl_2 + H_2 + 2NaOH$$

Then the chlorine reacts with water to form free chlorine:

$$Cl_2 + H_2O \rightarrow HOCl + H^+ + Cl^-$$

This equation does not reflect net pool behaviour once chlorine is consumed during normal sanitisation.

Outcome

Returning to the base concept of pH. pH is:

$$pH = -\log_{10}[H^+]$$

Pool water normally contains more dissolved CO_2 than would exist at equilibrium with air. This is deliberate and necessary. Excess CO_2 acts as a buffer, preventing pH from transforming every time water is disturbed by bathers splashing, wind ruffling the surface, or circulation-created disturbances.

In saltwater pools, the sustained pH rise is driven by the loss of this excess CO_2.

When a salt cell operates, it produces hydrogen gas. Those hydrogen bubbles do not create a pH rise on their own, but they increase physical aeration. This aeration speeds up the escape of dissolved CO_2 from the water. As CO_2 leaves, carbonic acid decreases and the pH rises.

This leads to an important distinction:

- **Core effect:** excess dissolved CO_2 in the water, determined primarily by total alkalinity (TA) and pH
- **Accelerating effect:** aeration caused by hydrogen gas bubbles from the salt cell

If the pool water does not contain excess CO_2. For example, when TA is low and/or the target pH is set higher, then additional turbulence has little effect because there is no surplus CO_2 available to out gas.

Chlorination itself, when fully accounted for, is effectively pH neutral. While salt cells produce hydroxide locally, acid formation balances this once chlorine dissolves and is later consumed through:

- sunlight (UV)
- oxidation of contaminants and nitrogen compounds.

In normal operation, these processes cancel, leaving no sustained pH rise because of chlorination chemistry alone.

In spotless or lightly used pools, a salt cell can briefly generate chlorine faster than it is consumed. During these periods, both free chlorine and hydroxide may rise slightly. This effect is temporary. Once the normal chlorine demand resumes, the system self-corrects.

One additional, and often overlooked, mechanism can produce a rapid pH rise.

If chlorine gas is not given sufficient time to dissolve (for example, because of very short pipe runs or poor return orientation), some chlorine gas may escape into the air instead of dissolving into the water. When this happens, the acidic component of chlorination is lost, while hydroxide remains, causing a large and sudden pH increase. This is typically resolved by improving chlorine dissolution, such as directing the nearest return downward to extend bubble contact time.

In summary, hydroxide accumulation does not cause the long-term pH creep seen in saltwater pools. Continuous CO_2 loss causes it, with hydrogen bubble aeration acting as an accelerator, not the source, of the problem.

Caution

This explanation of salt chlorination is incomplete. There are a huge number of reactions because of the electron generation of the salt chlorinator. This can include spinning off free radicals and several species of chemical compounds. The above explanation is intentionally simplified. For further information on the reactions possible, a chemistry textbook should be referenced.

ORP

Adding ORP to a salt system is not simple. As the above has explained, there are pH concerns; in addition, the chlorinator is also pumping out:

- Elevated chlorine levels
- Turbulence
- Free electrons
- Hydrogen gas

These add noise to an ORP signal. The primary issue is electron movement. ORP measures redox potential through electron exchange at the probe surface, so stray electrons and transient redox species can produce unstable or misleading readings.

Hydrogen gas generation and return line turbulence increase surface disruption, which speeds up carbon dioxide outgassing when excess dissolved CO_2 is present. In most pools and spas, dissolved CO_2 is intentionally held above atmospheric equilibrium because bicarbonate alkalinity is maintained for buffering and comfort. When CO_2 escapes, the pH rises. The resulting pH increase is therefore carbonate-driven, not a direct consequence of chlorine production. Because ORP is strongly pH dependent, this ongoing, chemistry-driven pH drift further undermines ORP stability as a control variable in salt systems.

Free radicals generated within the electrolytic environment may also interact with the ORP probe. As these species can exhibit a wide range of redox potentials, their presence may further contribute to erratic or non-representative ORP readings.

Different Principles of Operations

ORP and salt chlorination are coming at the same problem from different directions.

ORP is a near-instantaneous response to a change in chemical demand.

Salt chlorination is a continuous output device.

These are two different operational modes. Think something like heat pump = salt chlorinator and ORP/pH = gas heater. One method takes days to heat, while the other can heat in a few hours.

The two are simply different devices intended for different outcomes.

A better approach may be to mix the technologies separately.

Have the salt chlorinator provide the daily residual, say 2.0 mg/L, and the ORP/pH controller come in only if the residual dips below 2.0 mg/L.

Cyanuric Acid

Adding cyanuric acid is standard in Australian pools for sun protection of the chlorine.

ORP measures hypochlorous acid ions, as discussed previously.

Salt chlorination produces hypochlorous acid ions, as shown.

Cyanuric holds the hypochlorous acid, releasing only a small amount, between 3 and 10% as mentioned.

The salt chlorination reactions are generating stray electrons, hydrogen gas and hydroxide.

The result is an ORP/pH reaction chain that is indeterminate and is likely not practical as a control philosophy.

Before you say, but what about this or that water body where an auto controller is working with salt, stop as I did. Think about that specific pool. Everyone I thought of was running a gold-tip electrode with an alternative control, not ORP/pH.

CHAPTER 35

Chemicals Not to Use

Lists substances that interfere with probes or distort ORP readings.

ORP and pH probes are precision electrochemical instruments. Their sensing surfaces must remain clean, unobstructed, and chemically accessible. Anything that coats, plates onto, or interferes with these electrode surfaces will degrade measurement accuracy and dramatically shorten probe lifespan.

As outlined earlier in Chapter 23: Setting up the Site, several classes of substances foul or distort probe behaviour:

- Metals plate onto electrodes, altering ORP readings and causing drift.
- Organics (particularly sticky, high-molecular-weight compounds) form films that slow probe response.
- Proteins degrade into biofilm-like residues that clog reference junctions (part of the probe that completes the circuit).
- Polymers—natural or synthetic—adhere strongly to probe surfaces.
- Humic substances (brown, tea-stain organic matter) bind chemically to glass and platinum surfaces.

Although probes may appear to "work" temporarily when these substances are added to the pool, their service life and reliability are compromised. ORP controllers rely on fast, clean electron transfer at the sensing surface. Any contamination slows the response, distorts the millivolt reading, and causes the controller to make incorrect dosing decisions.

For this reason, chemicals based on metals, polymers, proteins, or heavy organics must not be used in any ORP/pH-controlled pool.

The following table lists common products, their categories, and the reasons they should be avoided.

Chemical Name / Product Type	Reasons to Avoid in ORP/pH-Controlled Pools
Aluminium Sulphate (Alum)	Aluminium is a metal; forms gelatinous floc that coats probes.
Poly Aluminium Chloride (PAC)	Contains aluminium (metal) and polymeric chains that adhere to sensor surfaces.
Poly Ammonium Chloride (Polydadmac, etc.)	Polymers form films on electrodes.
Copper-Based Algaecides	Copper is a metal; plates onto probe surfaces and shifts ORP.
Lanthanum-based Phosphate Removers	Lanthanum is a rare earth metal; can create fine precipitates that foul sensors.
Ferric-based Phosphate Removers (Iron products)	Iron is a metal; iron hydroxides coat and stain probe surfaces.
Polyquat Algaecides (Polyquaternium compounds)	Polymers that adhere tightly to ORP electrodes and interfere with electron transfer.
Enzyme-based Clarifiers	Proteins/organics which leave bio-films; break down slowly and produce probe coatings.
Natural "Bioclarifiers" and Plant Extract Flocculants	Organic and humic-like substances that stick to sensing elements.

Chemical Name / Product Type	Reasons to Avoid in ORP/pH-Controlled Pools
Flocculants containing Starch / Cellulose Derivatives	Natural polymers that gum up sensing membranes and diaphragms.
Metal-based Stain & Scale Products (iron, manganese blends)	Metals that can plate, discolour, or poison electrodes.
Calcium Chloride Clarifiers (rare specialty products)	Can cause mineral film deposition near probes in high-calcium water.
"Multifunction" Pool Shop Clarifiers (usually polymer blends)	Typically contain polymers, organics, or metal salts—all incompatible with ORP probes.
Phosphate-free clarifiers using acrylic or polyacrylamide	Synthetic polymers that coat electrodes before breaking down.
Metal-based algaecide blends (silver, zinc, copper combos)	Multiple metals increase plating and interference risks on electrodes.

CHAPTER 36

Heaters

Explains how heating systems affect chemistry and probe behaviour.

A lot of water bodies are using heating, and some are using cooling.

Heaters come in a variety of types. This is not a heating book, so I will only give surface treatment to heating. Specifically, we are only interested in the heaters from the point of view of ORP/pH.

Typical Gas Heater

A typical gas heater comprises a heating bar and a heat exchanger. The heat exchanger is what the water will pass through.

The heat exchanger is often made of copper or cupronickel. On each pass through the water, it will pick up ions of copper and will slowly deposit this into the water and plate out on the probe.

This type of heater is likely to shorten the life of the ORP/pH probe.

Other Heaters

Some older-style heaters may still be around. These heaters used iron heat exchangers. Similar to copper, the iron will be slowly added to the water on each pass.

This type of heater is not the best choice for an ORP/pH probe.

Heat Exchangers

Heat exchangers, external to the heater itself, are generally of two types.

Tube in tube, where the hot water circulates through one set of tubes and the cooler water circulates through another set of tubes. As the name states, heat is exchanged from the hotter side to the cooler side.

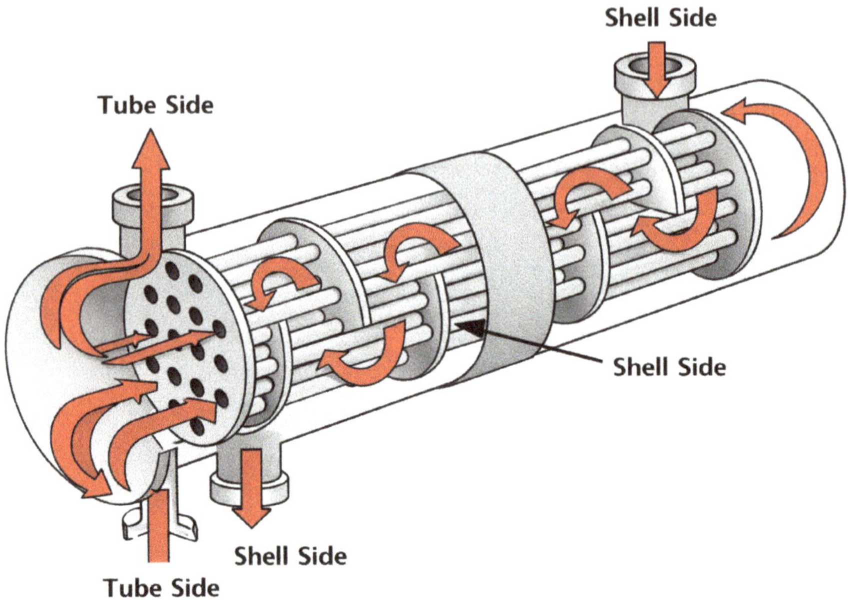

Plate heat exchangers: this type of heat exchanger comprises plates with hot water passing over one side and cooler water on the other side. Like a tube in tube, the heat is transferred from hot to cold.

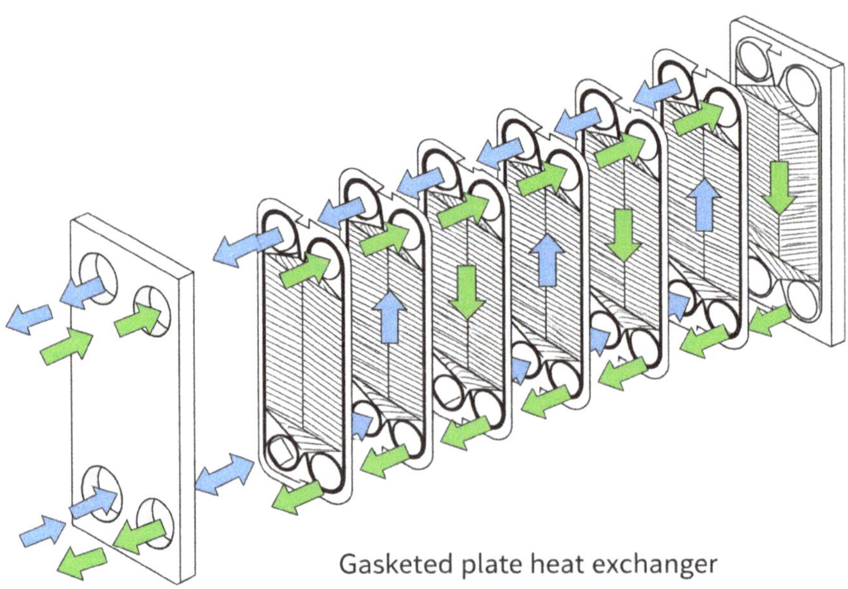

Gasketed plate heat exchanger

Both tube-in-tube and plate heat exchangers come in two distinct materials of construction:

- Stainless steel
- Titanium

For most chlorinated water bodies, it is more common to find titanium. Titanium is dealt with under heat pumps, below.

Heat Pumps

When heat pumps first appeared, they used stainless steel heat exchangers internally. However, because of constant corrosion leading to failures, they quickly adopted titanium.

In 1989 or so, I was using photometers for water testing in my pool shop. It was a time when heat pumps were first manufactured. The inventor would drop by with a water sample as the various heat exchanger prototypes failed. I recall saying I saw no choice but to use titanium.

Shortly after this, the polyethylene / titanium heat exchanger appeared and is still often found in heat pumps today.

For a little while, a heating company in Australia tried using cupronickel instead.

Today, I am not aware of any heat pump using anything other than titanium.

Titanium is chemically inert at the levels of chemicals encountered in recreational water bodies. Effectively, titanium can be ignored as a likely source of metal contamination for ORP/pH unless something truly extraordinary has happened.

Solar Tubes

Solar tubes tend to be of plastic construction. This should not be assumed, as there are instances where metals may be used. Asking is worthwhile.

Heating Issues

Metal contamination is one concern with heating. However, the heating itself adds issues as well. The obvious ones are:

- Chemical reaction speeds change
- Substances in the water dissolve faster
- Some chemicals tend towards more plating
- pH instability may increase
- CO2 evolution will alter
- Gas interchange will alter

The ORP/pH set and settled with a cooler temperature may suddenly change behaviour based on any of these changes.

Adding heating to a previously stable ORP/pH water body will probably require recalibration over several days as the temperature changes.

Spas Special case

Spas destroy ORP and pH probes faster because they combine every known probe-degrading condition:

1. High heat speeds up the chemical breakdown of glass membranes and reference junctions.
2. High bather load introduces oils, proteins, and organics, which foul sensors.
3. High sanitiser strength (especially bromine) and frequent shocking chemically attack electrodes.
4. Rapid pH swings strain probe chemistry.
5. High TDS and particulate loads clog reference junctions.
6. Small water volumes amplify chemical dosing errors, overcorrections, and fouling.
7. Turbulent, aerated flow damages sensors and causes unstable readings.
8. Frequent cleaning and recalibration, necessary in spas, shorten probe life.

All these mechanisms are individually harmful, and spas combine all of them at once.

CHAPTER 37

Make Up Water

Explains how incoming water changes chemistry and disrupts ORP control.

Water loss is a part of the normal management of any water body. Evaporation, bather splash-out, backwashing, and a raft of other everyday functions lead to water loss.

If there is a suspected leak, refer to my book, Pool and Spa Leak Investigation: A Practical Guide for Trained Professionals. This book offers practical guidance on isolating leaks.

Municipal vs. Bore vs. Dam Water

Working with the water treatment engineers, there is an expression they use that is applicable here.

The aquifer is subject to change without notice.

This applies to all filling water, whether it is municipal, bore, or from a dam or tank.

The water for any supply originates in several ways. These are broadly categorised as:

- Ground
- Surface
- Artificial

Groundwater typically shows water extracted from the earth. The usual method is via a bore pump.

Surface are typically dams, lakes and river-sourced waters.

Source	Bore	Dam	Desalination
pH Range	< 7.0	7.0–8.0	4.0–7.0
Salinity	High	Low	Lowest
Total Dissolved Solids	High	Low	Lowest
Metal Content	High	Varies	Lowest
Total Hardness	Higher	Moderate	Very Low
Alkalinity	Often High	Moderate	Very Low
Turbidity	Low – Moderate	High	Very Low
Organic Load	Low	High	Very Low
Ammonia / Nitrogen	Low	Variable	Very Low
Silica	High	Moderate	Very Low
Typical Temperature	Variable	Ambient	Ambient –Low
Corrosivity / LSI Tendency	Scaling	Balanced/ corrosive	Highly corrosive
Microbial Load	Low – Moderate	High	Very Low
DBP Formation Potential	Low	High	Very Low

Artificial water is water that is harvested from a source such as seawater and then subjected to processes. With seawater, the process is desalination, usually by reverse osmosis.

The table sets out these waters and their typical raw state:

Municipal water supply companies take this raw water and apply various treatments. Post-treatment, the water will differ from what is presented in the table. However, the raw characteristics colour the outcomes.

For example, desalinated water is usually treated with chemicals post-treatment to remineralize it.

In Perth, we regularly get a shandy of the three waters. This varies at the time of year and dam levels.

In Auckland, as a contrast, the water will usually be a dam-originated supply all year and usually more stable than Perth.

Knowing your area and its specific water make-up being added to the water body can assist with planning maintenance on your ORP/pH equipment.

Phosphate

Some main water supply companies will add phosphates to the incoming water to the pool. This is verifiable online for your area; if not, a phone call to the technical services of the water company is all that is needed.

Standard phosphate removal chemicals are generally not compatible with ORP/pH pools.

One option is to consider discussing with a drinking water filter company. They may have a product to assist.

Another option is to accept that the phosphates are present and run a slightly higher chlorine level to manage the water with them present.

Variation

When new water is added to the main water body, it will be:

- Lower in chemical levels
- Lighter

Than the existing water. Even if it is not added directly to the skimmer box or balance tank, it will probably end up passing into the pool plant in one slug.

An ORP/pH controller seeing this water will react immediately by adding the required chemicals the entire pool needs based on this new water. This can create significant overshooting depending on the controller's settings.

> **Rule:** *Sudden water changes create false readings and overdosing.*

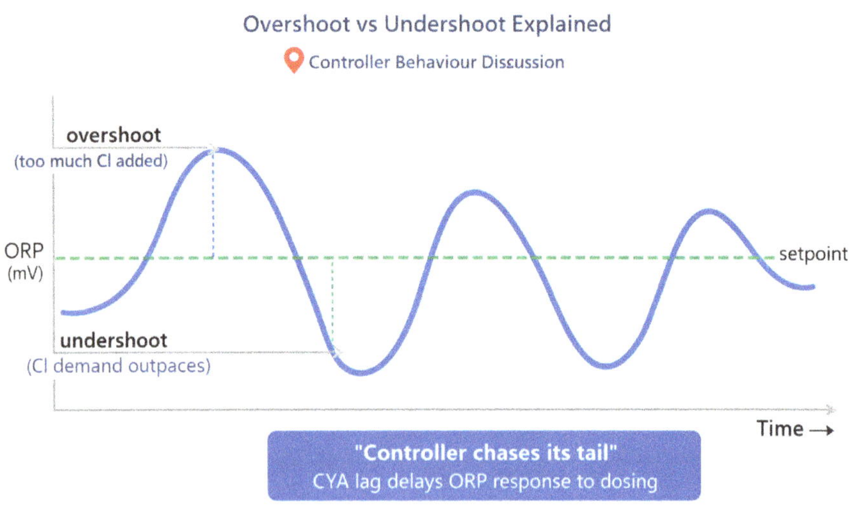

If you know a lot of new water is required, it is best to treat this similarly to manual chemical dosing. Isolate the probes and wait a minimum of one turnover before enabling them again.

Where the water top-up is automatic, it is best to discuss this with the chemical controller supplier and consider using a chemical controller with a greater change-accepting capacity available.

TDS Accumulation

The makeup of total dissolved solids is as important as the reading.

Total dissolved solids, primarily from chlorides, are less of a concern than the same reading supported primarily by phosphates and nitrates.

TDS accumulates in the water. Evaporation leaves the salts behind, and these build and build. These range in impact on the chlorine demand. Some chemicals are effectively inert and make little to no difference to the chemical used. Other components of the TDS can create extraordinary demand.

While it is not a strong marker, TDS should be considered if the water bodies' chemical consumption is elevated. This may be worth considering, and steps should be taken to reduce it.

Chemical Drift

The incoming water may have different parameter values than the main water body. Adding sufficient water may upset the chemical balance. The ORP/pH will change, and sometimes dramatically.

Always consider the incoming water and test it regularly to anticipate how it may affect the water body.

CHAPTER 38

Secondary Sanitisers

Explains how additional sanitising systems interact with ORP.

Some water bodies include secondary sanitisers. There is a wide variety of products in this area; however, I will focus on the main products:

- Ozone
- Ultraviolet
- Advanced Oxidation Processes

Ozone

As discussed, ozone (O3) has a powerful impact on ORP. Only a small amount of the O3 generated will dissolve into the water. The rest passes off as gas and needs to be destroyed.

Ozone has a residual in the water for a short time. It can be measured using DPD methods.

While it has a residual, it is not usually sufficient on its own as a complete sanitisation solution.

Ultraviolet

Ultraviolet is a different product. Effectively, the light is germicidal at 254 nm, and this is one of its functions. At a variety of dose rates, dose rates are measured in mJ/cm^2, the UV can destroy organics and may increase hypochlorous acid.

Flow rate, lamp age, quartz sleeve fouling, and water clarity all impact delivered dose and should be routinely monitored

Advanced Oxidation Processes

Advanced Oxidation Processes (AOPs)—typically UV combined with ozone or hydrogen peroxide—generate hydroxyl radicals (•OH). These radicals have extremely short lifetimes, often measured in microseconds, but they are among the most potent oxidisers available. Because of their short lifespan, AOP systems must generate radicals continuously and at the point of treatment. System design must account for reactor placement and ensure maximum contact with contaminants.

When Secondary Sanitisers Make ORP More Stable

Secondary disinfection may assist ORP stability by reducing the oxidation demand from organics, ammonia, and chloramines.

This reduction in oxidation demand can assist with the chlorine residual becoming more predictable, and this will assist the ORP/pH controller to have fewer oscillations caused by chlorine.

Both UV and AOP help stabilise the ORP, mostly by improving clarity and reducing organic load, while ozone will provide near-instant oxidation of pollutants.

By understanding these principles, it is possible to design safer, more efficient, and more stable sanitisation systems that integrate secondary technologies with ORP/pH control.

CHAPTER 39

Chemical Quality and Degradation

Explains how chemical strength changes over time and affects dosing performance.

This book would not be complete without mentioning chemical degradation.

A direct outcome of time and temperature, the rate of degradation will vary from place to place.

For this part of the chapter, I am going to concentrate on Western Australia as it is an excellent example of what we need.

Western Australia is a large state. For perspective, in land mass, it is 10 times larger than the United Kingdom, or about 3.6 times larger than Texas.

Transport

Rail is effectively limited to servicing a few larger towns, and some of the specialised mines have constructed their own rail for their own use.

Most of the transport of chemicals in this state is via road freight.

Temperature

Western Australia is one of the sunniest places in the world. We have more sunshine hours than any other Australian state. Perth is sunnier than most of California's population centres and significantly sunnier than the French Riviera.

Sunshine brings heat. Even on a cold, bleak day during winter, sitting in the car with the windows up can cause you to sweat with the sun's power coming in through the windows. This is because of the radiant heat from the sun. Cooler air temperatures do not change the radiant energy.

Imagine, instead of you sitting in the car with the windows up, it is a few kilograms of calcium hypochlorite or a drum of sodium hypochlorite. The product heats up.

This happens on trucks driving around the state delivering products. Even during winter, the product is heated up.

Sodium Hypochlorite

The worst product for degradation by heat is sodium hypochlorite. Once it reaches 40°C, it decomposes rapidly. However, trouble starts well before 40. Even in mid to high 20 degrees, the product gases off, and as it does, it loses strength.

> **Note:** Years ago. I had a Ford Escort van. I ran out of sodium hypochlorite in my shop. It was winter and raining, so I closed the shop and drove over to the suppliers to pick up 8 or 10 drums. As it was raining, I kept the windows up on the drive back. Even though it was not particularly warm and it was raining by the time I got back to my store, the chlorine had fumed enough to make me ill enough to close and go home.

When sodium hypochlorite is manufactured, it achieves a strength of 14-15%.

Even though the time from manufacturing to delivery to the chemical repackers is usually as short as the same or the following day, by the time they get it, the strength has reduced to 12% or thereabouts.

Over the following three months, the strength of the sodium hypochlorite reduces and typically stabilises at 4%. From 4%, the rate of decomposition slows; however, the final decomposition will see a drum of salty water in place of an active product.

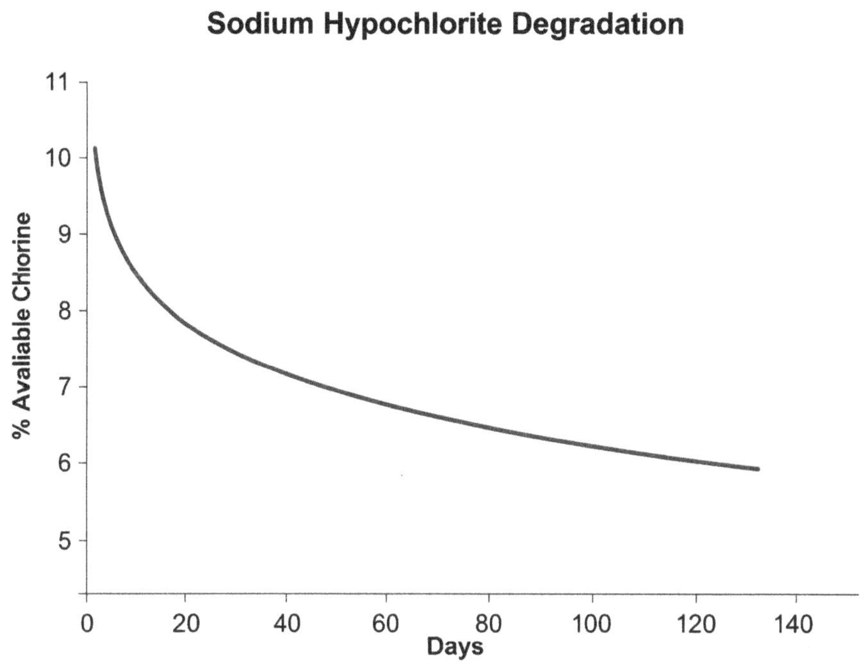

Heat increases this degradation. Not by a little, but by a lot.

Pay attention to how and where you store your sodium hypochlorite.

Keep chlorine cool (critical)

Store sodium hypochlorite < 25°C. If it is not possible,

- consider moving the chemical to an alternate location,
- storing a larger quantity, as a larger quantity will take longer to heat up,
- or perhaps providing a potential heat sink such as a water surround in the bund (upsize the bund as required).

Calcium Hypochlorite

Calcium hypochlorite degrades as well. Upon leaving the manufacturing plant, calcium hypochlorite will probably be at 70% strength. This strength declines rapidly and stabilises at around 65%.

Calcium hypochlorite does not degrade anywhere near as rapidly as sodium hypochlorite. However, it has another issue.

Some calcium hypochlorite will be white or creamy coloured; others, when you take the lid off, look a kind of grey colour. The grey is a possible indication of contamination. Before assuming the grey-coloured calcium hypochlorite is not good, check:

- Smell—Should smell of chlorine, not musty
- Texture—is the product mushy, or does it look fresh
- Appearance—any signs of rust flakes
- Adding to water—develops a brown residue, showing iron is present.

HCl Acid purity vs metals

Hydrochloric acid (HCl) can contain metallic impurities. Generally, this is with a less expensive product. Where metal (not stainless steel) storage tanks or piping have been used in the production. Common contaminants are copper, iron, and occasionally manganese.

There are a few issues with metallic contamination in the acid. Even in minute quantities, every dose can add a trace to the water.

It creates the potential for the water body to be stained.

- Yellow–brown staining (iron)
- Black or tea-coloured water (manganese)
- Turquoise deposits or surface discolouration (copper)

However, for ORP/pH controllers, it can:

- Create instability in the readings.
- Increase probe cleaning requirements.
- Decrease probe life.
- Increase chlorine demand.
- Increase total dissolved solids disproportionately.

Most people will blame "bad chlorine", "old chlorine", or "pipes", but many times, the stain source is simply cheap, metal-contaminated acid.

Signs of poor acid quality:

- Slight yellow tint in the drum.
- Brown sediment at the bottom of the drum.
- ORP dropping or oscillating abnormally after dosing acid.
- Stains are appearing even though the pH is well controlled.
- Chlorine demand is increasing without explanation.

Cheap acid saves cents per litre but introduces dollars of metal contamination into the system.

Sulphuric Acid Issues

Sulphuric acid (H_2SO_4) is used and sold in Western Australia. I am not sure about other parts of the world.

WA has some unique water properties that may protect the water bodies from the normal issues associated with sulphate. Specifically, the mains water here is low in sulphate and high in chlorides.

One advantage of this acid is that it does not introduce metals into the system. Sulphuric acids are cleaner than cheap HCl.

As mentioned, our water is likely providing some protection against the less favourable side effects of this product.

One factor I dislike about the sulphuric acid here is that the strength varies, which makes exact calculations difficult.

Calcium sulphate scaling

Sulphuric acid and hard or calcium-rich water can generate a calcium sulphate precipitate. This is a white, stubborn scale. It shows up predominantly on heaters and salt cells.

This is documented in industrial plating and cleaning cases. As mentioned, I have not seen this issue in WA.

Storage Temperature

Temperature has a direct, predictable influence on both the chemical environment and the electronics that measure ORP. Incorrect storage or heated environments can cause ORP readings to drift, destabilise, or become unreliable even when the water chemistry itself is correct.

This section explains why ORP changes with temperature, how heat affects probes, and how poor chemical storage conditions drive ORP instability in actual situations.

ORP temperature sensitive

The Nernst equation includes temperature as a term. As temperature increases,

- Reaction rates speed up
- Chlorine becomes more reactive
- Probe voltage output changes (≈ +2 mV per °C in typical conditions)

This is normal. The problem is not that temperature changes ORP; it becomes a problem when storage temperature amplifies these issues.

CHAPTER 40

High Storage Temperature Damages ORP Probes

Explains how storage conditions can reduce probe life and accuracy.

As mentioned, ORP sensors contain:

- A noble metal measuring electrode (platinum/gold)
- A reference electrolyte (often Ag/AgCl in KCl gel)

Heat can cause several issues, such as:

- Gel dehydration—the KCl gel slowly dries out.
- Internal pressure changes—If the pressure lowers enough, the water can contaminate the gel.
- Junction clogging.
- Reference drift.

These issues are just the effects on the actual probe.
An overheated probe will start:

- Being unstable
- Jumping around
- Slowing responses
- Unable to achieve a stable calibration
- Need a lot of corrections

This is indistinguishable from “bad chemistry” unless you know to check the probe history.

Why Storage Temperature Problems Look Like “Bad ORP Control”

In the field, ORP drift caused by temperature often appears as:

- ORP drops every afternoon
- ORP cannot rise after dosing
- ORP over corrects after acid injections
- ORP flatlines or oscillates
- ORP shows different numbers than manual testing

> **Note:** *ORP should increase as the afternoon ages —less sunshine = less chlorine loss = higher ORP.*

This can look like an issue with:

- Probe
- Controller
- Chlorinator
- Pool water

When the real root cause is chemical heat degradation or probe heat exposure.

CHAPTER 41

Practical Recommendations for Reliable ORP

Keep chlorine cool (critical)

Store sodium hypochlorite < 25°C if possible. One consideration could be a roof over the plant installation, with open sides to allow good airflow. This should help most of the time.

Better storage in WA and other parts of the world where summer temperatures exceed 35°C recommends that the pool plant be moved into the garage and some air cooling be added.

A plant room I worked on in the northern part of WA for a local council had air conditioning installed in the plant room. Temperatures in this part of the state are averaging 45°C in summer.

All parts of the world do not store drums in sheds without ventilation. This will lead to corrosion damage to any metal in the shed.

Keep probes cool

Never mount controllers on sun-facing walls. No electronic device can survive being heated in direct sunlight.

Any plant enclosure needs to be ventilated. A lot of the other equipment needs ventilation anyway; ideally, open louvres could be a reasonable choice on all sides.

If the probe is thought to have been stored hot or dry for any period, it is likely better to replace it than try to revive it.

Keep acids cool

Store acids away from metals, sunlight, and heat sources. The fumes can rapidly attack metals.

On a site we attended, the plant room was in the back corner of a car park, enclosed, but had some ventilation via a pair of louvre panels cut into the door. The door frame was metal. Even though the building was only 12 months old, the frame was rusting badly.

Radiant heat is also best avoided for acids.

Controller best practices

Chemical controllers are electronic items and may include heat sinks, fans, or other temperature management components internally. It is best practice to power them on and let them stabilise for about 20 minutes before enabling them to dose. This time allows the internal temperature to come into the normal operating range. A controller that is too hot or too cold will increase the likelihood of incorrect dosing.

Probe cables are signal cables. Signal cables should be kept away from power cables. Never cable-tie a signal cable to a pump motor power cable. Power cables induce a magnetic field around them. Cheaper signal cables may not be adequately shielded, and the magnetic field can distort the probe readings.

Always check the ORP value manually during the hottest part of the day.

CHAPTER 42

Troubleshooting Guide

Provides practical methods for diagnosing common ORP system problems.

Part of the issue with ORP/pH control can be simply working out what the likely cause of the issues being dealt with is.

> **Important diagnostic principle:**
>
> ORP failures are usually system failures, not instrument failures.
>
> Always assess hydraulics, pH stability, aeration, loading, and water quality before assuming probe or controller fault.
>
> **Rule:** ORP failures are system failures. Not controller failures.

Called out to the site, the controller is shooting all over the place, the numbers oscillating, perhaps stained swimwear around the pool, and the client looking at you. The pressure is on and real.

While this guide is for the ORP/pH controller, the principles apply to nearly all issues with any item of equipment.

Bad chemistry or bad equipment does not cause most failures in ORP and pH.

They are caused by moving faster than the water can respond.

Why Controllers Overshoot

ORP controllers respond to delayed information. The probe reports conditions that existed minutes or hours earlier, after transport and mixing delays. When conditions change, the controller reacts to this old signal and often adds more chemicals than needed. By the time the actual effect appears at the probe, excess dosing has already occurred. This overshoot is normal control lag, not a controller fault.

Identify the failure

What is the fault being dealt with?

- ORP reading is higher than expected
- ORP reading is unstable / oscillating
- ORP reading is flatlined / unresponsive
- ORP control is overshooting or undershooting
- ORP setpoint cannot be reached even with dosing maxed/ minimised

Environment Check

Before even touching the controller, consider other potential issues.

1. Is the pump basket clean and clear?
2. Does the filter need cleaning or backwash?
3. Are there chemicals in the drums?
4. Are the drums overflowing as water is bypassing the injector and back feeding into the drums?
5. Conduct a complete water test at the pool.
6. Conduct a complete water test at the probe (if possible)
7. Is the temperature normal for the pool and normal for the area?

8. Are there any signs of air entrapment within the plant?
9. Is there a flow restriction interfering with the water to the probe?
10. Any sign of manual dosing?
11. Check for indications of biofilm or debris in the sample lines.

There may be other items to check depending on the site and its specific equipment setup.

Only after checking thoroughly, look further into the instrumentation.

Check the probe

Is the probe:

- Clean.
- Tip is intact.
- Reference junction not blocked.
- Cable damaged.
- Probe not too old or expired.
- Installed in the correct position.
- Feeding lines are flowing correctly.
- No evidence of air or gas bubbles in the probe.
- Not installed after the chemical injection point.

Check the controller

This will vary from site to site and for different manufacturers' controllers.

For instance, on a site being regularly maintained, it is less likely that a connection is incorrect; it should be checked.

- Calibration date is known—if not, calibrate and add a date to the calibration log.
- Calibration is not overdue—if overdue, calibrate.

Remember: Calibration is a diagnostic step.
Not proof of safe control.

- Controller or dosing pumps are not in manual mode—diaphragm pumps often have a switch for external or internal signal. This should always be checked.
- Controller is not in alarm lockout.
- Controller is in the correct mode and not set to maintenance or some other setting.
- Setpoint seems realistic for the system.
- Hysteresis, bandwidth, time delays, time to dose, and similar settings look correct.

Remember: unless there is 100% certainty that a setting is incorrect, observe and do not adjust.

Turn off control

Disable the chemical feed. This may be done by simply putting a bucket of water in place of the chemicals. This helps simplify the system.

Wait and let the system stabilise.

Have you noticed that the enemy of ORP/pH control is being in a hurry? Waiting is often a sound method.

Now that the chemical feed is disabled, does the system stabilise?

If using the buckets of water, have one or both lost water into the system? (This can be a symptom of a jammed check valve in the injector or dosing pump.)

Manually correct water

Complete a water test and follow the steps laid out in Chapter 33, Manual Chemical Intervention, to correct the water balance, pH, and chlorine.

Do not move on without a stable water environment.

Fix the water, not the controller, first.

Re-enable the Controller and observe

Once the water is stable and known to be stable:

- Turn the controller back on.
- Watch the ORP and pH value for about 20 minutes. Are they stable or oscillating?
- Re-enable the chemical feed pumps.
- Watch the ORP and pH value for about 20 minutes. Are they stable or oscillating? Is the response markedly slow or perhaps too fast?
- Does the reading seem stable, or is there a sense of drifting?

Finalise

If, after the above steps, the control issue is still not stabilised, then consider replacing items in a sequential order:

- Probe cables
- pH Probe
- ORP Probe

Often, a fault can be only a problem with a probe cable.

An ORP probe typically only lasts a year or so.

Downloadable file:

https://drive.google.com/file/d/14LevUO5ZhMS6FUU1H3901ERzTAxUyToX/view?usp=sharing

CHAPTER 43

Choosing ORP Controller and Probes

Explains how to select equipment suited to real operating conditions.

Commonly, bad chemistry or bad controllers are, in fact, more likely an outcome of a poor choice of probe.

You are not choosing a probe for chemistry in a beaker; you are choosing a probe for an aerated, dirty, semi-open biological system.

Poor probe selection can lead to:

- Drift
- Slower responses
- False stability
- Constant calibration
- Premature replacement

Platinum Tip

Platinum tips are the most common for ORP electrodes; they are highly catalytic and very sensitive, which is why they are used. They are excellent for:

- Clean water
- Potable or industrial applications
- Well-controlled pools
- Maybe for controlling ozone or AOP systems

For many recreational water bodies, platinum is likely too good a catalyst. These probes are easily fouled and respond to transient oxidisers and stray redox events.

Platinum-tipped probes can literally see what we do not want seen, things like:

- Aeration
- Transient free radicals
- Chemical dosing

Platinum-tipped ORP probes make more sense in:

- Indoor commercial pools with low bather loads.
- Low bather load facilities.
- Where good pre-filtration is present.
- With no or minimal aeration.
- When using strict chemical discipline.

Gold Tip

For more real-world water bodies that do not exist inside a lab, a gold-tip ORP probe exists.

A gold tip probe is less catalytic and less reactive than a platinum-tipped probe, which provides a slower response time. Sometimes this type of probe is dismissed as being 'slower'; however, experience has shown that slower is likely better.

Gold-tip electrodes are:

- More stable in dirty, real-world water.
- Less sensitive to transient redox noise.
- Maybe more resistant to organic fouling, protein films, or body oils.

Often more usable in recreational water bodies.

Similar to platinum probes, the gold probes have some issues as well:

- Slower response times can be a negative.
- Can be less sensitive to subtle changes.
- Can under respond in very clean water.

Gold-tipped probes make sense for water bodies where:

- They are outdoors.
- Have higher bather loads.
- Subject to sunscreen, cosmetics, and the like.
- Stability matters more than a specific mV.

Platinum vs Gold Summary

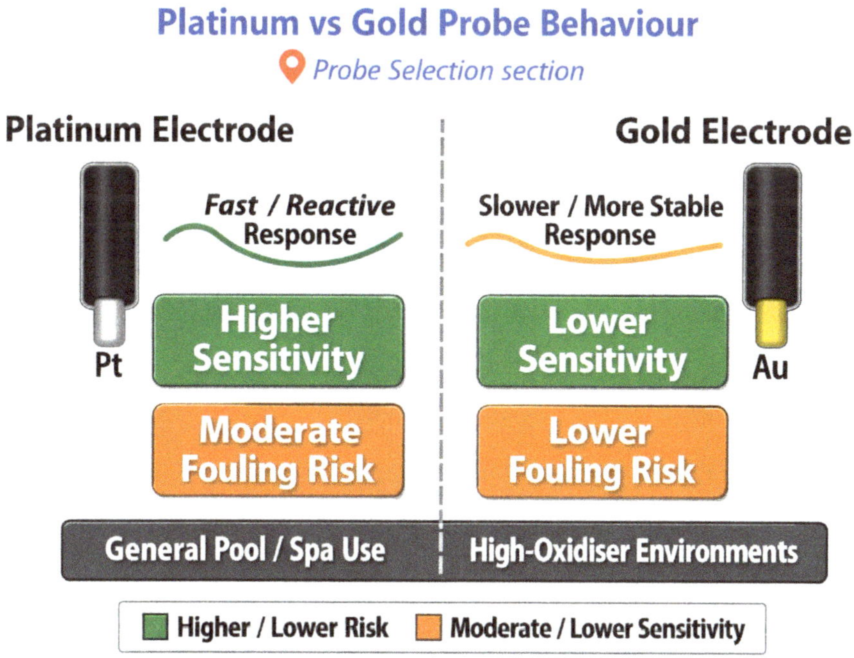

Platinum probes tell you everything. That does not mean platinum is better. It means they are more prone to noise and errors.

Gold-tipped probes tell you what matters. Sure, they are slower. But in average systems where turnover times are measured in hours, slower may in fact be an asset. Faster is not always better.

ORP probe sensing surface

The sensing surface on a probe comes in what is termed a flat or bulb surface.

Bulb tip ORP probes

This is the traditional rounded electrode design shape. This is very common in older designs.

They offer a small boundary layer of solution and give slightly better protection from hydraulic disturbances. However, they can create microzones where the readings are less accurate. They are also more prone to fouling and uneven cleaning.

This type of probe is commonly suggested as more suitable for:

- Stable industrial loops
- Low-turbulence environments
- Long-term immersion without frequent handling

Flat surface ORP probes

This is a smooth sensing surface flush with the probe body. It has advantages. They are easier to clean, less prone to creating microzones or boundary layers, and offer better response in flowing water.

This type of probe is more sensitive to hydraulics and requires better placement from a flow perspective. Always allow at least 5 pipe diameters before and after this probe to assist with smoothing to a laminar flow.

This type of probe will respond well to regular cleaning and maintenance. If this can be planned, the probes are well-suited for recreational water use.

Practical takeaway

Flat-surface probes survive real pools better if the hydraulics are correct.

Junction design

Gel-filled vs. open junction

One of those minor details that are often overlooked and are, in fact, critical.

Gel—filled reference

Often referred to as a sealed reference. This type of probe has an internal gel electrolyte. They allow slow ion exchange through a porous junction.

A gel-filled reference assists by allowing:

- Lower maintenance.
- Less likely to leak electrolyte.
- Good short-term stability.

However, these probes also allow the junction to clog easily with:

- Oils
- Polymers
- Proteins
- Scale

And once clogged, the response slows permanently. A clogged junction offers little opportunity to recover, culminating in a shorter life in real-world bathing water situations.

The best environment for this type of probe is clean water, where replacement is cheap and easy.

Open junction

An open junction can also be referred to as a flowing junction.

This is a liquid electrolyte that slowly exchanges with the sample. Some designs allow the refilling of the liquid.

This type of probe is much more resistant to clogging, leading to longer-term stability. These can usually be recovered from contamination and are more tolerant of dirty water.

However, they require more care than the gel-filled cousins; they can be mishandled and have a higher maintenance expectation.

This type of probe is better suited to:

- Commercial pools
- Outdoor pools
- High-organic environments
- Any ORP system expected to be reliable for years, not months

Key principle

In recreational waters, junction clogging kills probes faster than chemistry does.

No amount of calibration can make up for a badly selected probe installed in a less-than-ideal position.

pH Probes

pH probes fail for very similar reasons to ORP probes; however, they often fail faster. For real-world use, flat-surface glass with an open junction will often outperform traditional bulb instruments.

High temperature + aeration, + body oils can all lead to shortened probe life regardless of brand.

Suggested probe choice

Water body	Suggested probe type
Outdoor leisure	Gold, flat surface, open junction
Indoor lap pool	Platinum or gold, flat surface
Spa or high-aeration applications	Good luck with that
Commercial facility with strict chemical discipline	Platinum maybe possible
Domestic pool with enthusiastic pool shop chemistry	Gold likely survives

Choosing the wrong probe would be like trying to fit a bicycle wheel as a spare on a car. Not likely to work out well.

CHAPTER 44

Understanding Calibration

Explains what calibration does and what it does not fix.

What ORP calibration actually is

Calibration (aligning the probe to a known reference) is often misunderstood. Calibration does not make the ORP reading more accurate. It is simply ensuring that the probe and meter are reporting the same voltage.

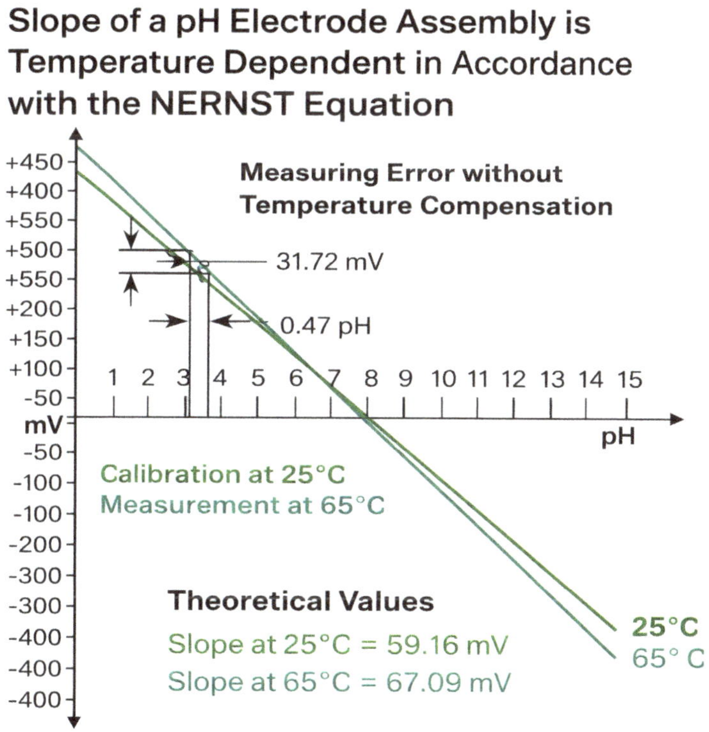

The probe will need correction over time for offset and slope. Slope is a term used a lot in chemical controllers. It is effectively saying the slope or angle of the plot on a graph at different mV with ORP. Over a period, this can change as the electronics in the unit and the reference in the probe age.

Effectively, it is simply a way of checking or setting that at, say:

- 255 mV, the electronics are reporting 255 mV, and
- 475 mV is really displaying as 475 mV

This is keeping the electronics accurate, not the water chemistry.

- Fouled
- Slow
- Or useless for control

Calibration is simply saying this meter, in this state of fouling, on this day, reads a specific millivoltage under a known condition. Nothing more, nothing less.

Calibration does not:

- make ORP "accurate"
- link ORP to mg/L chlorine
- compensate for cyanuric acid
- fix sluggish kinetics
- correct surface poisoning
- linearise a non-linear system
- turn a proxy into a measurement

Think of calibration as telling the meter how to correctly read the signal from the probe.

When ORP behaviour looks erratic, calibration often becomes the focus.

ORP failures are usually system failures, not instrument failures, and no amount of calibration can correct an unstable water environment.

Buffer solutions

Buffer solutions are simply buffers. With ORP, a solution manufactured by a laboratory to give a specific mV reading. Common mV buffers are ≈220 mV, ≈468 mV, and ≈ 650 mV.

Buffer solutions are also made at a pH. In pH, the common buffers are pH 4, pH 7, and pH 10.

> **Note:** *If calibrating pH, opt for pH 7 and pH 10 buffers. Water for bathing tends to be in the 7.2-7.8 range, and this span of buffers covers that range better than pH 4 to pH 7.*

The actual values, whether mV or pH, are less important; what is vitally important is the stability of the buffer.

- Always store buffers in a temperature-stable location. Leaving them in the truck's cab on a sunny afternoon is a sure-fire way to render the buffers useless.
- Check the expiry date. Every buffer will have an expiry date. At a pinch, an expired buffer may be better than nothing, but often it is not.

Renew your buffer solutions regularly. If calibrating a lot, it is worth diarising three or four weeks before the expiry date to order new buffers.

Calibration does not:

- *make ORP accurate in pool water*
- *correct fouling or slow response*
- *validate chemical safety*

Calibration still matters because it:

- *confirms that the probe electronics are functional*
- *reveals drift and aging*
- *acts as a diagnostic tool, not a guarantee*

Calibration frequency

The frequency of calibration is where the art lies. Calibration itself is a science. The frequency is more art.

Done too often, deterioration of the probe can be masked, yet leaving it too long can hide actual drift.

Instead of a calendar, try watching the readings over time and work on an acceptable/not acceptable drift.

Acceptance criteria

Large calibration offsets may indicate probe fouling or aging. Increasing offset over successive calibrations is a red flag that the probe needs replacement.

Slowing stabilisation in the buffer could show a junction or surface issue. Cleaning or replacement may be required.

Calibration can go well, with no issues, then over a few days, the values drift quickly, leading to a need to recalibrate. This is usually a sign of a poisoned probe and is likely a trigger to replace the probe.

Calibration offset

Not all controllers will show the offset amount as a number. However, if it does, record it in a log for this controller. Over time, this record can show that the probe is at the end of its life.

Some controllers will give errors like slope error or similar; these also indicate a likely issue with the probe. Cleaning is always the first step when these types of errors are encountered. If cleaning does not fix this, consideration should be given to replacement.

Lies, calibration lies

Calibration is important but not always sufficient.

A note on hysteresis, delay, and proportional settings

Many ORP controllers include a setting intended to slow or limit how aggressively the controller responds to changing readings.

Depending on the manufacturer, this may be labelled hysteresis, delay, deadband, percentage, XP, proportional band, or may not be exposed at all.

These settings do not change the accuracy of the ORP measurement and do not correct unstable water chemistry. They simply alter how the controller reacts to the probe signal it receives. Two controllers with identical real world behaviour may use different names for the same function or implement it internally with no adjustable parameters.

For this reason, this book does not define or recommend specific hysteresis values. In recreational water systems, apparent "hysteresis problems" are usually driven by transport delay, mixing time, pH drift, aeration, or probe condition. All factors belonging to the water body rather than the controller.

Treat these settings, where available, as response-shaping tools only. They cannot stabilise an unstable system, and no amount of controller change will compensate for chemistry or hydraulics changing faster than the water can respond.

CHAPTER 45

Operational Principles for ORP/pH Control

Sets out the core rules for using ORP control safely and effectively.

ORP is conditional, not unreliable.

From the incoming fill water to the last time the filter media was changed. ORP/pH can reflect the systems that support a water body. They are not operating in isolation.

Never interpret ORP without pH.

As has been shown in this book. pH is a critical factor for the ORP to be correct.

Any ORP number without context is just a voltage.

There is no dial in this number, and the pool will be okay. So many factors come into play to make any number vague at best.

Stable readings do not guarantee safe water.

Stable ORP/pH can be misunderstood as safe water. Nothing replaces a manual water test.

Calibration is a diagnostic tool, not proof.

Calibration is the verification of electronics. On its own, it is meaningless for water quality and not meant to be substituted for a manual water test.

If the water is unstable, turn the control off before correcting it .

Fix the water first. Then, only once the water is fixed and readings have become constant, enable automation.

Do not chase ORP faster than one turnover time.

Haste is not helpful. ORP/pH controllers cannot be adjusted faster than the turnover time of the water body. It takes one complete turnover to get an idea of the impact the changes made are having.

Some water bodies will never suit ORP control.

Reasonably easy to identify, these are water bodies with high aeration and unstable pH. There may be others under heavier than usual use, or have some peculiarity in the filling water. Not every water body is suited to ORP/pH.

CHAPTER 46

Alternative chemical controllers

Introduces other control methods and where they may be more suitable.

ORP measures the environment, not concentration.

This does not make ORP weak as a control method. It makes it different from concentration-based control systems.

When concentration measurements become necessary, a different sensor class is used.

Realistically, for recreational water use, the main alternative controllers are based on:

- Amperometric (measures actual chlorine directly)
- Colourmetric (uses colour change to measure chlorine)

The problem with these alternatives is that the cost of the controllers is usually about three to four times the cost of an ORP/pH controller. Besides the controller cost, they are also more commonly packaged with diaphragm dosing pumps, and these can often be five times more expensive than a peristaltic pump.

However, there are in fact several technologies. Many of these are not currently used in our market.

Method	Measures	Pros	Cons	Typical Use
ORP	Oxidation tendency	Cheap, simple	Indirect, unstable in pools	Domestic pools
Amperometric	HOCl (free chlorine)	Fast, accurate	Needs stable flow and pH	Commercial pools
Colorimetric (DPD)	Free/total chlorine	Most accurate	Reagent cost	Compliance, municipal
ISE	Chloride / TRO	Robust	Not HOCl-specific	Industrial
UV Absorbance	Indirect oxidant	No reagents	Not selective	Treatment plants
Optical/Luminescent	HOCl impact on dye	Stable	New tech	High-end potable
Timed Feeding	None	Simple	Not adaptive	Spas, splash pads

CHAPTER 47

Industry Standards and Legal Responsibilities

Explains regulatory expectations and operator responsibilities for water safety.

Throughout this book, the examples are often drawn from Western Australia. That is where I have spent most of my career, and WA is a useful laboratory: extreme UV, road-freighted chemicals degrading in the heat, and a regulatory framework that has been quietly progressive in several areas. But water chemistry does not respect state borders, and neither do the problems this book is trying to solve.

This chapter maps the major standards frameworks that a reader in Australia, the United Kingdom, the United States, or Europe might encounter. It is not a substitute for reading the documents themselves or for taking professional legal advice. Standards change. Enforcement varies. Always verify the current version in your jurisdiction.

A word of caution before diving in: standards are minimum floors, not performance targets. A pool that just meets the minimum free chlorine requirement on a busy summer afternoon may still produce unsafe water if the pH is drifting, the CYA is high, and the ORP is sitting below 600 mV. The numbers in standards documents are the beginning of the conversation, not the end.

The International Baseline: WHO Guidelines

The World Health Organization's Guidelines for Safe Recreational Water Environments, Volume 2: Swimming Pools and Similar Environments (WHO, 2006) is the closest thing the industry has to a global reference point. It underpins or is referenced by regulatory frameworks from Perth to London to São Paulo.

The WHO document recommends a free chlorine residual of 1 mg/L for a conventional public pool with good hydraulics and filtration operating within its design bather load. For hot tubs, the WHO suggests a higher range of 2 to 3 mg/L, given heavier bather loads and elevated temperatures. For pools using supplementary UV or ozone, a lower residual of 0.5 mg/L may be acceptable.

On pH, the WHO recommends maintaining the range 7.2 to 7.8 for chlorine-based disinfection, and 7.2 to 8.0 for bromine-based and other non-chlorine processes. For ORP specifically, the 2006 guidelines reference 720 mV as a value associated with good microbial condition — a figure that has been widely cited (and sometimes misunderstood) in the years since.

> **Note:** *The WHO ORP figure of 720 mV is a guideline for microbial safety, not a controller setpoint. As this book has explained at length, reaching 720 mV on the probe is not a guarantee of water safety, and the reading can be achieved through mechanisms entirely unrelated to chlorine. Use it as a reference; do not treat it as a target to be dialled in.*

The WHO guidelines do not have legislative force in most jurisdictions. Their value is as a scientific foundation. When a national or state standard is disputed, the WHO document is usually the first reference both sides reach for.

Australia: The WA DOH Code of Practice and National Context

In Western Australia, public aquatic facilities are regulated under the Health (Aquatic Facilities) Regulations 2007, which empowers the mandatory Code of Practice for the Design, Construction, Operation,

Management and Maintenance of Aquatic Facilities, published by the WA Department of Health (DOH). The Code was most recently updated in March 2024.

The WA Code addresses water quality parameters, including free chlorine, combined chlorine, pH, cyanuric acid, total alkalinity, and calcium hardness. Turnover rates and filtration requirements are specified. For commercial pools, the average standard turnover rate in WA is 2.5 hours, tighter than in many other jurisdictions. Reflecting both the high bather loads and elevated UV intensity typical of the WA climate.

On cyanuric acid, WA allows up to 50 mg/L, which is higher than in many comparable jurisdictions. This is deliberate accommodation for the UV environment. Pools in WA that do not use CYA will lose chlorine to sunlight at a rate considered astonishing in many other parts of the world, including the other states of Australia, where outdoor pools are rarer and UV levels are more modest. Chemical controller manufacturers in WA have adapted their default settings to reflect this.

The WA Code operates within the broader Australian context. Other states have their own regulatory instruments — NSW Health publishes its own Public Swimming Pool and Spa Pool Advisory Document, South Australia has its Standard for the Operation of Swimming Pools and Spa Pools, and so on. There is no single national Australian pool standard, which creates variability. The Australian and New Zealand Standard AS/NZS 1646 covers equipment for water treatment, and AS 1345 governs pipe identification. Operators working across state lines will need to check requirements in each jurisdiction.

> **Note:** *Domestic backyard pools in Australia are governed differently from public facilities. In WA, the Health (Aquatic Facilities) Regulations 2007 do not apply to residential pools. Barrier and fencing requirements fall under the Building Regulations. Operators working in the residential space are not subject to the same chemical testing and record-keeping obligations as commercial operators, but the water chemistry does not know that.*

United Kingdom: PWTAG and HSG282

The United Kingdom has no single piece of legislation specifically governing swimming pool water chemistry. Instead, operators navigate a combination of the Health and Safety at Work Act, the Management of Health and Safety at Work Regulations, and guidance documents — of which the two most important are HSG282 and the PWTAG Code of Practice.

HSG282

HSG282 — Managing Health and Safety in Swimming Pools — is published by the Health and Safety Executive (HSE) and covers spa pools in particular detail. HSG282 is the document that inspectors reach for when attending a spa pool incident. It addresses Legionella risk, water temperature management, and disinfection requirements for hot water environments. A key HSG282 requirement is the maintenance of a measurable residual disinfectant at all times, with specific attention to Pseudomonas aeruginosa as an indicator organism.

For spa pools, HSG282 recommends a free chlorine residual of 3 to 5 mg/L, and total bromine of 4 to 6 mg/L if bromine is used. These are higher than the equivalent pool values and reflect the additional challenges of hot, heavily aerated, high-bather-load environments, exactly the conditions that this book has argued make ORP/pH control unreliable.

PWTAG

The Pool Water Treatment Advisory Group (PWTAG) is a non-commercial UK membership organisation that publishes the Code of Practice for the Management and Treatment of Swimming Pool Water and the Swimming Pool Water book. The HSE recognises PWTAG guidance as an authoritative resource for pool operators, and enforcement authorities typically consider PWTAG guidance when assessing whether a pool's operating procedures are adequate.

PWTAG recommends a free chlorine range of 1 to 3 mg/L for pools using chlorine-based disinfection, with an upper limit of 3 mg/L.

Combined chlorine should be below 50% of the free chlorine reading and never exceed 1 mg/L. The preferred pH range is 7.2 to 7.4 — notably tighter than the WHO range. PWTAG has consistently emphasised that a lower pH within the acceptable range maximises disinfection efficiency, a position this book has explored from an electrochemical standpoint.

On ORP, PWTAG's Technical Note on Residual Control provides a frank appraisal of both redox and amperometric measurement. It notes that a redox probe responds to any oxidiser in the water, not only chlorine, and that the signal is pH-sensitive. The technical note observes that some controllers display a calculated free chlorine value derived from ORP and pH readings — a point worth keeping in mind when reading a display that appears to show ppm rather than mV. PWTAG does not mandate specific ORP setpoints for general pool use, but the document is clear that ORP alone cannot be relied upon as confirmation of adequate disinfection.

PWTAG technical notes are updated periodically and are available on the PWTAG website. Practitioners in the UK should check for current technical notes before relying on any specific parameter.

> **Note:** *PWTAG guidance is best practice, not legislation. This distinction matters: a pool that meets PWTAG guidelines has a strong defence in the event of a public health incident or enforcement action, but the absence of a PWTAG-compliant procedure is not, by itself, a legal violation. What is required legally is that operators comply with health and safety law, which PWTAG guidance helps to interpret.*

United States: MAHC, PHTA, and State Codes

The United States does not have a single federal standard for recreational water chemistry. Regulation falls to individual states and localities, which means there are fifty-plus different legal frameworks operating simultaneously. The result is considerable variation: a commercial pool in Florida operates under different rules than the same pool in California or Minnesota.

The CDC Model Aquatic Health Code (MAHC)

The CDC's Model Aquatic Health Code, now in its 5th edition (2024), is the most comprehensive single reference document available to US operators. It is model guidance — states and localities can voluntarily adopt it, in whole or in part, to save time developing their own codes. Several jurisdictions have done so.

The 2024 MAHC recommends a minimum free chlorine of 1 mg/L for pools and 3 mg/L for hot tubs and spas. For pools using cyanuric acid, the minimum rises to 2 mg/L. The recommended pH range is 7.0 to 7.8. The MAHC is notable for triggering immediate closure when free chlorine exceeds 10 mg/L while bathers are present — a hard upper limit reflecting a genuine safety concern, not just a water chemistry preference.

On ORP, the 2024 MAHC does not mandate a specific ORP setpoint for all pools, but acknowledges ORP measurement as part of automated chemical control systems. Some of the US state codes go further: Florida, for instance, specifies that where ORP controllers are installed in public pools, the measured potential must be kept between 700 and 850 mV. Several other states — including Arizona, Colorado, Ohio, and Wyoming — reference ORP in their pool codes as a supplementary parameter. Research published in peer-reviewed literature has suggested that 650 mV be considered a minimum standard (the value associated with near-instantaneous kill of E. coli and poliovirus inactivation), though this is not universally codified.

PHTA (Pool & Hot Tub Alliance)

The Pool & Hot Tub Alliance (PHTA) is the main US industry body and the ANSI-accredited standards developer for residential and public pools and spas. Its most relevant instruments include ANSI/PHTA-11 (Water Quality for Public Pools and Spas) and the 2023 ANSI/PHTA/ICC-2 Standard for Public Pool and Spa Operations and Maintenance, developed jointly with the International Code Council.

For residential pools, PHTA guidance has historically recommended a free chlorine range of 2.0 to 4.0 mg/L — a wider and higher band than WHO or PWTAG recommendations. This reflects the US market's greater reliance on cyanuric acid in outdoor pools: with

CYA present, a higher nominal chlorine level is needed to deliver the same concentration of active HOCl. The relationship between CYA, chlorine, and ORP discussed at length in this book is accounted for in that higher recommended range, even if it is not always explained that way to the end customer.

NSF International publishes NSF/ANSI 50, which covers equipment and chemicals for swimming pools. Equipment certified to NSF/ANSI 50, including ORP probes and chemical dosing systems, must meet the performance requirements set out in that standard. This is relevant when specifying equipment: NSF/ANSI 50 certification is a meaningful verification of performance, and enforcement authorities in the US often require it.

> **Note:** *Because US pool regulation is state level, an operator or technician working across multiple states will encounter meaningful differences in testing frequency requirements, permitted CYA levels, chemical approval lists, and automation provisions. The MAHC is a useful harmonisation reference, but it is not law unless the jurisdiction has adopted it. Always verify local requirements.*

Europe: DIN 19643 and Beyond

Germany's DIN 19643 standard for swimming pool water treatment is notable for two reasons. First, it is technically rigorous, more so than most comparable documents in other jurisdictions. Second, Germany has enforced an ORP standard for public pools since 1982, making it one of the longest-running real-world experiments in ORP-based regulation.

DIN 19643-1 (most recently revised in 2023) sets a free chlorine range of 0.3 to 0.6 mg/L for conventional pools — significantly lower than UK, Australian, or US figures. This reflects the German approach of relying on engineered treatment systems rather than high residuals: DIN 19643 pools are expected to have high-quality filtration, flocculation, and often ozone or UV as a primary treatment, with chlorine providing a smaller residual sanitiser role. Chloramine levels are limited to 0.2 mg/L, which is tight.

For whirlpools, the DIN standard permits 0.7 to 1.0 mg/L. The ORP range cited in DIN 19643 is 700 to 770 mV, though as noted in this book (and as Richard Falk's work shows), the absolute millivolt reading varies between probe manufacturers and is better used as a process control reference than a safety guarantee.

Elsewhere in Europe, national standards vary. France, the Netherlands, and the Scandinavian countries all have regulatory frameworks, some based on DIN 19643, others on national legislation or WHO guidance. The EU does not have a single binding directive on swimming pool water quality — the EU Bathing Water Directive applies to natural bathing sites, not artificial pools. European operators should consult national requirements directly.

Standards at a Glance

The table below summarises key parameters across the major frameworks. Values are indicative and subject to revision; always confirm current versions.

Table note: Spa/hot tub values refer to high-temperature spa pools unless otherwise noted. CYA presence affects effective chlorine requirements — refer to source documents for CYA-adjusted figures. ORP references are for guidance; probe-to-probe variation means absolute mV values are not directly comparable across instruments.

Where the Standards Agree

Across all these frameworks, a few things are consistent, and they map directly onto the themes of this book.

The pH must be maintained within a defined range. Every standard agrees on this. The specific range varies slightly — DIN 19643 allows a lower floor of 6.5 in some configurations, while CDC and WHO prefer a floor of 7.0 — but the principle is universal: pH outside the acceptable range compromises disinfection and bather comfort.

Manual testing cannot be eliminated by automation. No standard anywhere permits an automated controller to substitute entirely for manual verification.

Framework	Free Cl (mg/L) — Pool	Free Cl (mg/L) — Spa	pH Range	ORP Reference	Status
WHO 2006	1.0 (min)	2.0–3.0	7.2–7.8	720 mV	Guideline
WA DOH Code (2024)	Per Code	Per Code	7.2–7.8	Not mandated	Regulatory (WA)
PWTAG (UK)	1.0–3.0	3.0–5.0	7.2–7.4	No fixed setpoint	Best practice / HSE-recognised
HSG282 (UK spas)	N/A	3.0–5.0 Cl 4.0–6.0 Br.	7.0–7.8	Not mandated	HSE guidance
CDC MAHC 2024 (US)	1.0 (min)	3.0 (min)	7.0–7.8	Varies by state	Model code (voluntary)
Florida pool code	Per state code	Per state code	7.2–7.8	700–850 mV (if an ORP controller is installed)	Regulatory (FL)
DIN 19643 (Germany)	0.3–0.6	0.7–1.0	6.5–7.6	700–770 mV	National standard (DE)

A few years ago, I trialled running a colorimetric analyser and an amperometric controller alongside manual testing on one pool. I was hopeful I could get sufficient data to show a potential for reduced manual testing from 3 times daily to perhaps 1 or even less. I could not get the three systems to stabilise and agree. Manual testing and amperometric matched close enough for the purpose; however, the colorimetric just would not settle. I tried a few alternative positions and so on; however, for whatever reason, it would not stabilise sufficiently. After six months, I abandoned this trial.

The CDC recommends testing at least twice per day for public pools, and more frequently under heavy use. PWTAG specifies daily testing. Australian codes require regular manual testing regardless of whether automated controls are in place. This is not bureaucratic caution; it is a recognition of exactly the problem this book has explored: that ORP/pH controllers can read correctly while the water is unsafe, and can read incorrectly while the water is fine.

Secondary sanitisers reduce residual requirements in some frameworks. Both PWTAG and the WHO acknowledge that UV or ozone supplementation can support reduced chlorine residuals. DIN 19643 is structured around this principle. This is an area where Australian practice is still catching up, though the 2024 WA Code amendments reflect developing thinking.

Where the Standards Diverge — and Why It Matters for ORP

The differences between frameworks are as informative as the similarities.

Germany's low free chlorine range (0.3 to 0.6 mg/L) and simultaneous ORP requirement (700 to 770 mV) are significant data points. It reflects a deliberate design philosophy: invest in treatment infrastructure — filtration, flocculation, ozone — and use a small chlorine residual with continuous ORP monitoring to confirm that the system is working. In this model, ORP is checking the output of a well-engineered system, not compensating for deficiencies in one.

At 0.5 mg/L free chlorine in an unstabilised German pool at pH 7.0, the ORP will readily reach 700 mV. At 2.0 mg/L free chlorine in an Australian pool with 50 mg/L CYA at pH 7.6, the ORP may struggle to reach 580 mV. Both pools meet their respective national standards. Only one of them is giving the ORP controller a workable signal. Therefore, transplanting ORP setpoints between regulatory environments without change is a recipe for problems.

The US state-by-state variation means that an ORP requirement that applies in Florida does not apply in California. Practitioners working nationally or internationally need to develop the habit of checking local requirements rather than assuming a setpoint learned in one jurisdiction transfers cleanly.

CYA thresholds vary significantly. WA's 50 mg/L upper limit reflects the UV environment. The CDC MAHC suggests 90 mg/L as a maximum under its current guidance, though this is disputed in some quarters. Germany effectively prohibits CYA use in public pools covered by DIN 19643. The UK discourages it in commercial pools. These differences have direct consequences for ORP: at the same free chlorine level, a pool with 50 mg/L CYA will show a meaningfully lower ORP than one with no CYA, as discussed in Chapter 20. Where Operational Rules Finally Break Down.

Legal Responsibilities of the Operator

Regardless of jurisdiction, the legal and moral responsibility for maintaining safe water rests with the operator. A controller that malfunctions is not a defence; the operator's duty of care is not delegated to the equipment. This is worth understanding in plain terms.

In Australia, the Health (Aquatic Facilities) Regulations 2007 in WA specify what must be tested, how often, and what records must be kept. Local government environmental health officers enforce regulations and have powers of entry and closure. Failure to maintain required chemical levels or failure to maintain the records that show compliance can cause improvement notices, prohibition orders, and, in serious cases, prosecution.

In the UK, the duty of care falls under health and safety law. The Management of Health and Safety at Work Regulations 1999 require employers to assess and manage risks, which for pool operators means having and following safe operating procedures that address water quality. An incident at a pool that cannot show it was following its own procedures — or that had no written procedures — will face scrutiny under HSE investigation.

In the US, enforcement is local. State health departments inspect public pools and have the authority to close them. Civil liability for water-quality-related illness is also significant: plaintiffs' attorneys are well aware of the MAHC and state codes, and showing that an operator deviated from standard practice is a meaningful avenue of litigation.

For operators fitting or relying on ORP/pH controllers, the key legal point is this: in every jurisdiction covered here, automated chemical control does not substitute for manual testing and record-keeping. Wherever you are in the world, you are expected to test manually, record the results, and be able to show that the water was within acceptable parameters at the time the tests were conducted. An ORP controller display is not a record.

> **Note:** *This chapter is not legal advice. Regulations change, enforcement practices vary, and specific situations require professional legal advice. The intent here is to give practitioners a working map of the landscape, not a legal opinion.*

Testing Frequency: A Practical Summary

Testing frequency requirements vary, but the following represents broadly accepted commercial practice across jurisdictions. Always check and apply your specific local requirements.

A Note on Staying Current

As noted in Chapter 48 of this book, new research continues to refine our understanding of ORP in recreational water. The standards frameworks described here are living documents, not monuments. The WA Code of Practice was updated in 2024. The CDC MAHC reached its 5th edition in 2024. DIN 19643 was revised in 2023. PWTAG publishes technical notes ongoing.

What changes is often the fine detail — specific parameter ranges, equipment certification requirements, testing protocols, but occasionally the fundamentals shift as well. The treatment of cyanuric acid in commercial pools has moved in several jurisdictions over the past decade. Secondary disinfection requirements are developing. ORP's role in regulatory frameworks is not static.

The references below are the primary sources used in this chapter. Each organisation publishes its guidance documents online, and most provide update notifications for subscribers. Getting on those mailing lists is, frankly, worth more than reading any secondary summary of the standards, including this one.

Parameter	Minimum Frequency (Commercial Pool)	Notes
Free chlorine / total bromine	At least twice daily; hourly in heavy use	Manual DPD test. Controller display is not a substitute.
pH	At least twice daily	Manual test each time free chlorine is tested.
Combined chlorine / chloramines	Daily	Indicator of bather load impact and ventilation adequacy.
Total alkalinity	Weekly	Affects pH stability and CO_2 behaviour.
Calcium hardness	Weekly to fortnightly	LSI balance; scale and corrosion risk.
Cyanuric acid	Monthly (outdoor pools)	Accumulates; affects ORP behaviour significantly.
TDS	Monthly	Accumulation indicator; consider dilution when elevated.
Microbiological (bacteria)	Monthly (lab sample)	PWTAG/DOH requirement for commercial pools. Not operator-testable by kit.
ORP (mV)	Continuous (if controller installed); logged regularly	Controller data. Always reconcile with manual DPD test.

Key References and Resources

- WHO Guidelines for Safe Recreational Water Environments, Volume 2: Swimming Pools and Similar Environments (2006) — apps.who.int
- WA Department of Health — Code of Practice for the Design, Construction, Operation, Management and Maintenance of Aquatic Facilities (updated March 2024) — health.wa.gov.au
- PWTAG Code of Practice for the Management and Treatment of Swimming Pool Water — pwtag.org

- HSE — Managing Health and Safety in Swimming Pools (HSG179) and Hot Tubs: Managing Health and Safety Risks (HSG282) — hse.gov.uk
- CDC Model Aquatic Health Code, 5th Edition (2024) — cdc.gov/model-aquatic-health-code
- PHTA/ANSI Standards — phta.org/standards
- NSF/ANSI 50 — Equipment and Chemicals for Swimming Pools, Spas, Hot Tubs and Other Recreational Water Facilities (current edition) — nsf.org
- DIN 19643-1:2023 — Treatment of Water of Swimming Pools and Baths, Part 1: General Requirements — din.de

CHAPTER 48

Summing Up

Reinforces the key ideas and practical takeaways from the book.

I started this book somewhat sceptical of the advantages of ORP/pH control.

It is not going away soon, and to be fair, it is a cost-effective technology that works broadly in many situations.

In considering the contents and findings overall, an ORP/pH pool with the best chance of being stable and reliable will have:

- Incoming water filtered for metals
- Additional phosphate filtration if required
- Minimal surface disturbance or aeration
- Good to excellent circulation
- Well-maintained filtration
- A laminar pipe location free of air for probe placement
- Appropriate choice of probe for measurement
- Strict limits on the chemicals used

If these factors are all met, ORP/pH can work with reasonable success most of the time.

For water bodies that cannot meet these requirements, and if the budget allows, it is worth considering one of the alternative chemical-control technologies.

It is also important to recognise that even in 2026, researchers continue to publish new findings that overturn assumptions from the 2000s.

A truth that was "correct" 15 years ago can be wrong today because:

- Probes changed
- Water treatment changed
- Bather loads changed
- CYA norms changed
- Metals and clarifiers changed
- Modelling got better

Based on everything uncovered during the writing of this book, I am still sceptical of many of the claims surrounding ORP/pH, but perhaps it can work reliably in the right water body.

CHAPTER 49

ORP, pH and Chlorine, Real World Example and Reference Tables

Provides reference data for comparing ORP readings to measured chemistry.

Worked example—controller pool drifting low ORP

- Pool volume: 80,000 L
- Free chlorine measured: 0.6 mg/L
- pH: 7.8
- CYA: 40 mg/L
- ORP: 610 mV

Operator action:

Add 1 L of sodium hypochlorite (12.5%)

Expected result:

Free chlorine rises ≈ 1.5 mg/L

ORP increase modest (not proportionally) because:

- CYA buffers HOCl
- pH still high
- probe response delayed

Observed result:

ORP rises to 640 mV after 30 minutes

Interpretation:

Controller behaving normally

Next step:

Lower the pH slightly to stabilise ORP response

Tables

These values are compiled from field observations and published controller behaviour datasets rather than theoretical predictions alone. They are provided as a guide and should not be used as a precise predictor.

Any chemical change, even as minor as having a different total alkalinity or calcium hardness level, can affect these readings.

Reference tables for ORP, pH, and chlorine levels, with and without cyanuric acid (CYA).

ORP for Unstabilised Water (0 ppm CYA)

Free Chlorine (mg/L)	pH 7.2	pH 7.4	pH 7.6	pH 7.8
0.5	~660 mV	~640 mV	~620 mV	~600 mV
1.0	~700 mV	~680 mV	~660 mV	~640 mV
2.0	~740 mV	~720 mV	~700 mV	~680 mV
3.0	~770 mV	~750 mV	~730 mV	~710 mV

Table – ORP for Stabilized Water (50 ppm CYA)

Free Chlorine (mg/L)	pH 7.2	pH 7.4	pH 7.6	pH 7.8
0.5	~540 mV	~530 mV	~520 mV	~510 mV
1.0	~580 mV	~570 mV	~560 mV	~550 mV
2.0	~620 mV	~610 mV	~600 mV	~590 mV
3.0	~650 mV	~640 mV	~630 mV	~620 mV

CHAPTER 50

What the Controller Doesn't Tell You

There are no shortcuts in ORP control.

If you are looking for a number to set and walk away, this system will eventually prove you wrong.

ORP does not behave according to neat rules or fixed targets.

It behaves according to the system it is a part of.

What follows are not instructions.

They are observations of how real pools behave.

Ignore them, and the system will eventually remind you.

The System Comes First

A pool is not a tank of water.

It is a system made up of hydraulics, mixing, filtration, chemistry, temperature, and load.

ORP sits inside that system.

It does not control it.

It reflects it.

When the system is stable, ORP tends to be stable.

When the system is unstable, ORP becomes unstable.

Nothing the controller does can override that.

ORP Does Not Measure What You Think It Measures

ORP is not chlorine.

It is not a direct measure of sanitiser.

It is a voltage.

That voltage reflects the combined behaviour of everything happening at the probe tip:

pH

oxidisers

reducers

temperature

probe condition

and local water conditions at that exact point

A stable number does not guarantee safe water.

An unstable number does not always show a lack of chlorine.

Without context, an ORP reading is just a number.

pH Drives ORP Behaviour

ORP follows pH.

When the pH rises, ORP falls.

When pH falls, ORP rises.

This happens regardless of the chlorine level.

It is not something the controller can distinguish.

If the pH is moving, ORP will move with it.

Attempts to control ORP without first stabilising pH will usually result in overshoot, undershoot, or constant change.

Operators often respond to ORP movement by changing chlorine setpoints.

Many times, the underlying cause is pH.

pH Follows CO_2

pH movement is not random.

It follows the movement of carbon dioxide.

When water is disturbed:

- CO_2 leaves the water
- pH rises
- When CO_2 accumulates:
- pH falls

Every source of disturbance:

- aeration
- bathers
- turbulence
- heating
- wind

contributes to this process.

This is continuous.

It does not stop when the plant room is quiet.

The controller is responding to a system that is always moving.

Disturbance Propagates Through the System

Disturbances in a pool do not stay local.

It moves through the system.

Poor mixing leads to uneven chemistry. And unstable chemistry means the probe is guessing.

Uneven chemistry leads to unstable pH.

Unstable pH leads to unstable ORP.

By the time the controller reacts, it is responding to the result of multiple upstream effects.

Therefore, the same pool can behave differently from one day to the next with no obvious change at the controller.

Clean Water Produces Clear Signals

ORP probes rely on clean, exposed surfaces to function properly.

Anything that coats or interferes with the probe affects the reading:

- oils
- organics
- polymers
- metals

A fouled probe does not fail obviously.

It usually produces a suppressed or sluggish signal.

The controller responds to that signal as if it were correct.

This often results in overdosing.

From the controller's perspective, nothing unusual has happened.

It received a low number and acted on it.

The Controller Does Not Understand the Pool

The controller compares one number to another.

If the reading is below the setpoint, it doses.

If the reading is above the setpoint, it stops.

That is the entire decision.

It does not know:

- what the actual chlorine level is
- whether the probe is clean
- whether pH has drifted

- whether the chemical is degraded
- whether fresh water has just been added
- how many people are in the pool

These things look the same to the controller.

They all appear as a number.

Delay Is Built Into the System

Chemicals do not act instantly.

Water does not mix instantly.

Transport time exists between:

- injection point
- circulation path
- probe location

The controller doses based on what the probe reads now, not what the system will look like in five minutes.

This delay is a common cause of overshooting.

By the time the probe reading rises, additional chemicals are already in transit.

The system catches up after the fact.

High Stability Conditions Mask Problems

In calm systems:

- large volume
- low disturbance
- consistent load

ORP control can appear very effective.

Numbers remain stable.

Dosing appears smooth.

In these conditions, the controller looks intelligent.

In reality, the system is simply not challenging it.

When conditions change:

- heavy load
- weather shift
- aeration increase

The behaviour changes with it.

This can happen suddenly, even after long periods of apparent stability.

Difficult Systems Expose Limitations

Small volumes, high aeration, and rapid change create conditions where ORP control struggles.

Spas, spray pads, and highly aerated pools are typical examples.

In these environments:

- pH moves quickly
- CO_2 movement is constant
- mixing is uneven
- demand changes rapidly

The controller is always behind the system.

The result is oscillation, overshoot, and instability.

This is not a failure of the device.

It is a mismatch between the control method and the system.

Numbers Without Context Are Misleading

An ORP value on its own does not describe water safety.

The same reading can represent very different conditions depending on:

- pH
- cyanuric acid
- temperature
- chemical mix

Two pools can have the same free chlorine and very different ORP values.

Two pools can have the same ORP and very different sanitising conditions.

What matters is not the number itself, but what is driving it.

Stability Must Be Created, Not Controlled

Controllers do not stabilise pools.

Stability comes from:

- proper hydraulics
- effective mixing
- correct filtration
- appropriate chemistry
- realistic loading

Once stability exists, the controller can help maintain it.

Without that foundation, the controller amplifies instability.

The Role of the Operator

The controller is a tool.

It is not an authority.

Manual testing is not there to confirm the controller.

It is there to confirm the water.

The operator's role is to:

- understand the system
- recognise instability
- interpret behaviour
- and intervene when required

No controller replaces that.

Final Observation

ORP behaves consistently.

But it does not behave simply.

It reflects the system exactly as it is, not as we expect it to be.

When the behaviour seems wrong, it is usually because something upstream is changing.

Understanding that relationship allows the system to be managed.

Not the number on the display.

CHAPTER 51

Appendix Chemical Potential E° (V)

Lists standard electrode potentials for reference and comparison.

Name	Formula / Couple	E° (V)	Role
Acetic acid (rare, cleaning)	CH_3COOH	N/A	Organic acid; no pool redox role.
Algaecide, copper-based	$Cu^{2+} + 2e^- \rightarrow Cu$	+0.34	Non-oxidizing biocide; stains if mismanaged.
Algaecide, polyquat	(Cationic polymer)	N/A	Non-oxidizing; surfactant biocide.
Alkalinity (as bicarbonate)	HCO_3^-	N/A	Main buffering species.
Aluminium (III)	$Al^{3+} + 3e^- \rightarrow Al$	−1.66	Coagulant precursor; not an oxidiser.
Aluminium sulphate (alum)	$Al_2(SO_4)_3$	N/A	Hydrolyses to Al $(OH)_3$ floc; no redox.

Name	Formula / Couple	E° (V)	Role
Ammonia (contaminant)	NH_3/NH_4^+	N/A	Reacts with HOCl → chloramines; no direct E°.
Ammonium chloride	NH_4Cl	N/A	Source of ammonia in testing/experiments.
Ascorbic acid (stain remover)	$C_6H_8O_6$	N/A	Reducing agent for iron stains (no standard aqueous E° cited here).
Bicarbonate	HCO_3^-	N/A	Alkalinity buffer.
Boric acid / Borates	$B(OH)_3$ / $B(OH)_4^-$	N/A	pH buffer; mild algaestat.
Bromide	Br^-	N/A	Precursor/endpoint for bromine systems.
Bromine	$Br_2 + 2e^- \rightarrow 2Br^-$	+1.07	Spa oxidiser/sanitiser.
Calcium chloride	$CaCl_2$	N/A	Raise calcium hardness.

Name	Formula / Couple	E° (V)	Role
Calcium hypochlorite (as chlorine source)	$Ca(OCl)_2$	See Cl_2/Cl^- couple	Solid chlorine donor; oxidiser strength per Cl_2 couple.
Calcium carbonate	$CaCO_3$	N/A	Scale/LSI component; no redox.
Carbonate	CO_3^{2-}	N/A	Buffer; scale former with Ca^{2+}.
Carbonic acid (dissolved CO_2)	H_2CO_3	N/A	pH/alkalinity system.
Chloramines (mono/di/tri)	NH_2Cl / $NHCl_2$ / NCl_3	N/A	Disinfection byproducts; odour/irritant; no single E°.
Chlorate	ClO_3^-	N/A	End-product/impurity; non-oxidizing in pools.
Chloride	Cl^-	N/A	Endpoint of chlorine oxidation; spectator ion.
Chlorine (free)	$Cl_2 + 2e^- \rightarrow 2Cl^-$	+1.36	Primary oxidiser/sanitiser.

Name	Formula / Couple	E° (V)	Role
Chlorine dioxide (if used)	$ClO_2 + e^- \rightarrow ClO_2^-$	≈ +0.95 to +1.0	Specialty oxidiser for problem water.
Chlorite	ClO_2^-	N/A	Intermediate /byproduct; not a pool oxidiser.
Cyanuric acid (CYA)	$C_3H_3N_3O_3$	N/A	Stabiliser only; no redox behaviour.
Dichlor (Na salt)	NaDCC (Na-dichloroisocyanurate)	N/A	Chlorine donor + CYA; oxidiser strength from HOCl formed.
Dioxide, chlorine	(see "Chlorine dioxide")	—	See above.
Ferric/ferrous iron	$Fe^{3+} + e^- \rightarrow Fe^{2+}$	+0.77	Redox pair; staining, oxidation by chlorine.
Ferrous iron to metal	$Fe^{2+} + 2e^- \rightarrow Fe$	−0.44	Precip. /Staining; reducing side reference.
Fluorine (reference)	$F_2 + 2e^- \rightarrow 2F^-$	+2.87	Benchmark only.
Hydrochloric acid	HCl ($\rightarrow H^+ + Cl^-$)	N/A	pH ↓; the Cl^- is fully reduced (no oxidiser action).

Name	Formula / Couple	E° (V)	Role
Hydrogen peroxide	$H_2O_2 + 2H^+ + 2e^- \rightarrow 2H_2O$	+1.78	Strong oxidiser; cleanup/auxiliary.
Hydroxide	OH^-	N/A	pH ↑; no redox.
Hypobromous acid / hypobromite	$HOBr$ / OBr^-	(see Br_2 couple)	Bromine system speciation: oxidiser strength follows Br_2/Br^-.
Hypochlorous acid / hypochlorite	$HOCl$ / OCl^-	(see Cl_2 couple; also, $ClO^- + H_2O + 2e^- \rightarrow Cl^- + 2OH^-$ = +0.90)	Active chlorine species; pH-dependent strength.
Iodine (some spa systems)	$I_2 + 2e^- \rightarrow 2I^-$	+0.54	Weaker oxidiser; sometimes used with oxidizing assist.
Lanthanum chloride	$LaCl_3$	N/A	Phosphate remover; precipitation chemistry; no redox.

Name	Formula / Couple	E° (V)	Role
Manganese (permanganate test/cleanup)	$MnO_4^- + 8H^+ + 5e^- \rightarrow Mn^{2+} + 4H_2O$	+1.49	Occasional specialty cleanups.
Monopersulphate (MPS) *(operates via persulphate/per sulphate - radical)*	HSO_5^- (→ $S_2O_8^{2-}$ couple shown above)	+2.01 (for $S_2O_8^{2-}/SO_4^{2-}$)	Non-chlorine shock oxidiser.
Nitrate	NO_3^-	N/A	Fully oxidised N; nutrient; no oxidiser role in pools.
Nitrite (rare, problem indicator)	NO_2^-	N/A	Can consume oxidiser; not a pool oxidiser itself.
Oxygen (acidic couple)	$O_2 + 4H^+ + 4e^- \rightarrow 2H_2O$	+1.23	Aeration background; weak vs chlorine.
Ozone	$O_3 + 2H^+ + 2e^- \rightarrow O_2 + H_2O$	+2.07	Strongest practical oxidiser with persulphate / peroxide.
Peracetate / peracetic acid (rare in pools)	CH_3CO_3H (various couples)	N/A	Sometimes in industrial water, it is not common in pools.

Name	Formula / Couple	E° (V)	Role
Peroxide (hydrogen)	$H_2O_2 + 2H^+ + 2e^- \rightarrow 2H_2O$	+1.78	See above.
Persulphate	$S_2O_8^{2-} + 2e^- \rightarrow 2SO_4^{2-}$	+2.01	See above.
Phosphate	PO_4^{3-}	N/A	Nutrient; precipitated by La^{3+}; no redox.
Polyquat	—	N/A	Non-oxidizing polymeric algaecide.
Potassium monopersulphate	$2KHSO_5$ (MPS blend)	(see $S_2O_8^{2-}$ couple)	Non-chlorine shock ingredient.
Sodium bicarbonate	$NaHCO_3$	N/A	Raises alkalinity; no redox.
Sodium bisulphate	$NaHSO_4$	N/A	Dry acid; pH ↓; no redox.
Sodium carbonate (soda ash)	Na_2CO_3	N/A	pH ↑ / alkalinity ↑; no redox.
Sodium chloride (salt)	NaCl	N/A	Salt for SWCGs; Cl^- is fully reduced.
Sodium hydroxide	NaOH	N/A	pH ↑; no redox.

Name	Formula / Couple	E° (V)	Role
Sodium hypochlorite (liquid chlorine)	$NaOCl$	(see Cl_2/Cl^- couple)	Donor of $HOCl/OCl^-$; oxidiser strength per chlorine.
Sodium percarbonate (if used)	$2\ Na_2CO_3 \cdot 3\ H_2O_2$	(see H_2O_2 couple)	Releases hydrogen peroxide; non-chlorine oxidiser.
Sodium thiosulphate (dechlorinator)	$S_2O_3^{2-}$ (reducing agent)	N/A (acts as reductant)	Reduces chlorine to chloride; no standard E° listed here.
Sulphate	SO_4^{2-}	N/A	Fully oxidised; spectator ion.
Sulfuric acid	H_2SO_4	N/A	pH ↓ (commercial plant rooms); no oxidiser role in pool concentrations.
Trichlor (TCCA)	$C_3N_3O_3Cl_3$	N/A	Strongly acidic chlorine donor; oxidiser strength via HOCl after hydrolysis.

Name	Formula / Couple	E° (V)	Role
Urea (bather waste)	$(NH_2)_2CO$	N/A	Consumes oxidiser; forms DBPs; no E°.
Zinc (anodes in some systems)	$Zn^{2+} + 2e^- \rightarrow Zn$	−0.76	Sacrificial anode; corrosion control.

For clarity, while the E° (V) helps understand the impact of the chemicals, this is not a direct impact on the ORP. Adding 1 mole of sodium hypochlorite to our bucket of deionised water will not add 0.9 or 1.0 to the ORP.

CHAPTER 52

Appendix Painting Plastics

Provides a practical method for preparing and painting plastic components.

Painting plastic is a somewhat frustrating task. Paint is applied, looks great, dries, and then, as soon as it is touched, picked up, or moved, the paint literally falls off in the affected section.

This happens because plastic has a layer of oils on the surface known as plasticisers. These plasticisers prevent the paint from binding to the plastic; instead, the paint binds to the plasticiser, and as this is effectively an oily coating, the paint is effectively binding to nothing.

To successfully paint plastic, the plasticisers need to be removed. There are three ways around this that I am aware of:

1. **Heat—**plastics manufacturers use flaming before printing to clear the plasticisers
2. **Special undercoats—**Hardware stores sell undercoats that claim to prime plastic to prevent this issue.
3. **Chemical—**Use of solvents can remove the plasticiser and provide a suitable surface.

Heat can be used either as a flame or with a heat gun. The technique is to effectively heat the surface sufficiently to allow the plasticisers to float off and prepare the surface. Plastic should be painted while warm.

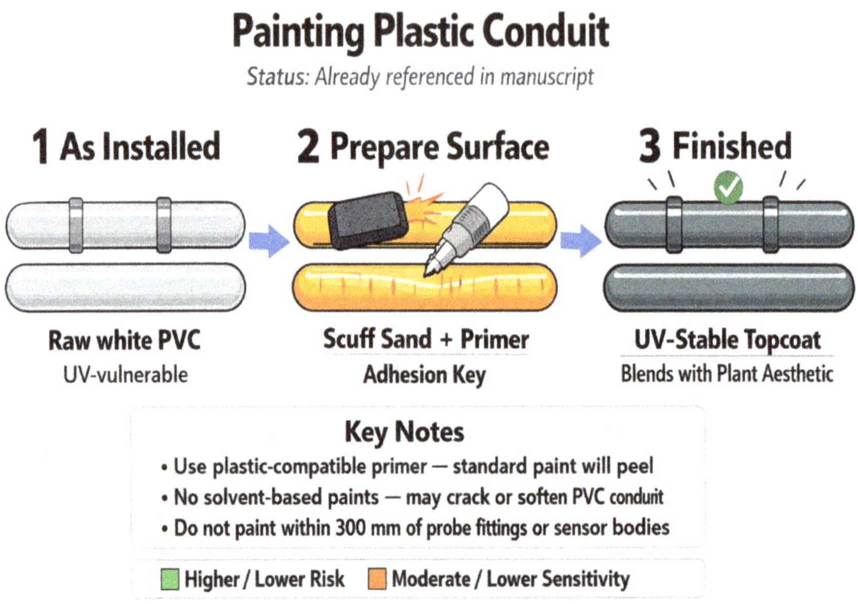

Aesthetics + UV protection — convert any source imagery to original line art, remove trademarks

Special undercoats are likely expensive. However, this may be the best option depending on location and local regulations, safety requirements, and the like. As with any chemical used on site, always get and keep a copy of the safety data sheet and follow this document's safety requirements.

Our current method is to use methyl ethyl ketone (MEK). It is essential to obtain, keep, and read the safety data sheet for this product. Best used outside in a well-ventilated area. Don the personal protective equipment as listed in the safety data sheet. Literally wipe the conduit with the MEK, and paint is applied immediately. MEK is one component of PVC glue. The surface of the plastic becomes sticky when wiped, and this is what the paint is applied to. This method literally glues the paint to the plastic. Once the paint is dry, it is a lot more resistant to scrapes and scuffs. Another coat over the top, and it is going to last for years, not days.

For small-bore conduit, we normally paint it before fixing it in place. Sure, conduit can be painted once mounted, but it is fiddly and somewhat difficult.

CHAPTER 53

About the Author

David Watson

David Watson has spent over forty years working with recreational and public water systems across multiple regulatory environments, focusing on the practical reality of keeping bathers safe in real, imperfect pools. His work spans domestic and large-scale commercial facilities and includes hands on responsibility for water chemistry, hydraulics, microbial risk management, and regulatory compliance.

Over his career, David has worked at the intersection of field practice and applied chemistry, developing a reputation for diagnosing problems that persist after standard explanations fail. His background includes dangerous goods consulting in Western Australia, technical advisory work on bulk chemical storage and dosing systems, and involvement as a joint inventor on a US-filed water chemistry dosing process.

David's engagement with chlorine chemistry, redox behaviour, and ORP is not academic in the abstract, but grounded in decades of observing how theory survives—or breaks—when exposed to bather load, heat, aeration, degraded chemicals, and imperfect infrastructure. His thinking has been informed by direct engagement with researchers in the field, including detailed technical correspondence on chlorine speciation and redox kinetics.

He entered the aquatic industry by accident, stayed because the problems would not leave him alone, and has remained driven by a simple belief: a pool is not a static container of water, but a dynamic system whose safety depends on understanding how its parts actually interact.

Known among colleagues as methodical, bookish, and defiantly persistent in a quiet way, David has a reputation for returning to unresolved technical questions long after others have moved on, not out of stubbornness for its own sake, but because the system itself has not yet given a satisfactory answer.

David is less interested in products or folklore than in causal chains that can be tested, explained, and trusted under pressure. His writing reflects that disposition. This book exists because too much practical knowledge about ORP and pH control remains either misunderstood or oversimplified, and because the consequences of getting it wrong are real.

This guide is written for technicians, operators, and managers who work in the field, from someone who still thinks like a technician first—and who has spent a lifetime learning where the theory ends and responsibility begins.

Contact and Support Information

For further assistance, technical support, or to provide feedback on this guide, please visit:

Website: www.davidwatsonpoolconsulting.com

Here you can find:

- Additional resources and downloadable forms
- Contact details for consulting and training services
- Updates and announcements

If you require direct support, please use the contact options provided on the website to reach out to David Watson Pool Consulting.